Imagined Futures in Science, Technology and Society

Imagining, forecasting and predicting the future is an inextricable and increasingly important part of the present. States, organizations and individuals have to almost continuously make decisions about future actions, financial investments or technological innovation without much knowledge of exactly what will happen in the future. Science and technology play a crucial role in this collective attempt to make sense of the future. Technological developments such as nanotechnology, robotics or solar energy shape how we dream and think about the future, while economic forecasts, gene tests or climate change projections help us to make images of what might occur in the future.

This book provides one of the first interdisciplinary assessments of how scientific and technological imaginations matter in the formation of human, ecological and societal futures. Rooted in different disciplines such as sociology, philosophy and science and technology studies, it explores how various actors such as scientists, companies or states imagine the future to be and act upon that imagination. Bringing together case studies from different regions around the globe, including the electrification of the German car infrastructure, or genetically modified crops in India, *Imagined Futures in Science, Technology and Society* shows how science and technology create novel forms of imagination, thereby opening horizons towards alternative futures. By developing central aspects of the current debate on how scientific imagination and future-making interact, this timely volume provides a fresh look at the complex interrelationships between science, technology and society.

This book will be of interest to postgraduate students interested in science and technology studies, the history and philosophy of science, sociology, cultural studies, anthropology, political sciences, future studies and literary sciences.

Gert Verschraegen is Associate Professor in the Department of Sociology at the University of Antwerp, Belgium.

Frédéric Vandermoere is Associate Professor in the Department of Sociology at the University of Antwerp, Belgium.

Luc Braeckmans is Professor in Philosophy at the Department of Philosophy of the University of Antwerp, Belgium.

Barbara Segaert holds a Master's Diploma in Oriental Studies, Islamic Studies and Arab Philology from KU Leuven, Belgium, and a Master's Degree in the Social Sciences from the Open University, UK.

Routledge Studies in Science, Technology and Society

21 **Nuclear Disaster at Fukushima Daiichi**
Social, Political and Environmental Issues
Edited by Richard Hindmarsh

22 **Internet and Emotions**
Edited by Tova Benski and Eran Fisher

23 **Critique, Social Media and the Information Society**
Edited by Christian Fuchs and Marisol Sandoval

24 **Commodified Bodies**
Organ Transplantation and the Organ Trade
Oliver Decker

25 **Information Communication Technology and Social Transformation**
A Social and Historical Perspective
Hugh F. Cline

26 **Visualization in the Age of Computerization**
Edited by Annamaria Carusi, Aud Sissel Hoel, Timothy Webmoor and Steve Woolgar

27 **The Leisure Commons**
A Spatial History of Web 2.0
Payal Arora

28 **Transparency and Surveillance as Sociotechnical Accountability**
A House of Mirrors
Edited by Deborah G. Johnson and Priscilla M. Regan

29 **The Fukushima Effect**
A New Geopolitical Terrain
Edited by Richard Hindmarsh and Rebecca Priestley

30 **Digitizing Identities**
Doing Identity in a Networked World
Edited by Irma van der Ploeg and Jason Pridmore

31 **The Globalization of American Infrastructure**
The Shipping Container and Freight Transportation
Matthew Heins

32 **Science, Risk, and Policy**
Andrew J. Knight

33 **The Ethics of Ordinary Technology**
Michel Puech

34 **Imagined Futures in Science, Technology and Society**
Edited by Gert Verschraegen, Frédéric Vandermoere, Luc Braeckmans and Barbara Segaert

Imagined Futures in Science, Technology and Society

Edited by
Gert Verschraegen,
Frédéric Vandermoere,
Luc Braeckmans and
Barbara Segaert

Routledge
Taylor & Francis Group
LONDON AND NEW YORK

First published 2017
by Routledge
2 Park Square, Milton Park, Abingdon, Oxon OX14 4RN

and by Routledge
605 Third Avenue, New York, NY 10017

Routledge is an imprint of the Taylor & Francis Group, an informa business

British Library Cataloguing in Publication Data
A catalogue record for this book is available from the British Library

Library of Congress Cataloging in Publication Data
A catalog record for this book has been requested

ISBN: 978-1-138-21737-9 (hbk)
ISBN: 978-1-315-44084-2 (ebk)

Typeset in Times New Roman
by Wearset Ltd, Boldon, Tyne and Wear

Contents

Illustrations

Figures

Tables

Contributors

Jan Baedke is Assistant Professor in the Department of Philosophy I and the 'Spaces of Anthropological Knowledge' Mercator Research Group, Ruhr University, Bochum. His research interests include the history and philosophy of the life sciences and philosophical anthropology. He received his PhD in 2014. His thesis assesses explanatory novelties in modern biology and uncovers inter- and intradisciplinary conflicts hindering the theoretical integration of epigenetics. He has published in international journals such as *Studies in History and Philosophy of Biological and Biomedical Sciences* and *History and Philosophy of the Life Sciences*. He co-edited the volume *Anthropologie 2.0? Neuere Ansätze einer philosophischen Anthropologie im Zeitalter der Biowissenschaften* (LIT, 2015).

Luc Braeckmans is Professor in Philosophy at the University of Antwerp. He is Director of Academic Affairs of the University Centre Saint-Ignatius Antwerp.

Matthias Gross is Professor of Environmental Sociology at the University of Jena and holds a joint appointment with the Helmholtz Centre for Environmental Research – UFZ, Leipzig, Germany, where he is head of the Department of Urban and Environmental Sociology. His areas of expertise are: alternative energy systems, environmental sociology, ignorance and the knowledge society, real-world experiments and environmental innovation, science and technology studies and theories of nature–culture relations. Among his recent and forthcoming publications are: *Green European: Environmental Behaviour and Attitudes in Europe in a Historical and Cross-Cultural Comparative Perspective* (Routledge, co-edited with Audrone Telesiene) and *Handbook of Energy and Society* (Oxford University Press, co-edited with Debra Davidson).

Kim Hendrickx is a postdoctoral fellow of the Research Foundation – Flanders. He studied philosophy and anthropology in Leuven and obtained his PhD at the University of Liège. He is interested in both science and technology, as particular practices, and their relation to the human body. More particularly, he studies the mediations that connect science, technology and specific conceptions of the human body, health and disease. He is a founding member of the Belgian STS Network and he was a visiting scholar at Maastricht University (2011–2013) and a postdoctoral researcher at the University of Leuven.

Kaat Louckx is a postdoctoral fellow at the Department of Sociology of the University of Chicago (BAEF Fellowship) and at the Center for Social Theory of Ghent University. Her research interests include history of science, sociology of knowledge and social theory. Her work appeared, among others, in *The Sociological Review*, *Social Science History* and *Nations and Nationalism*.

Andreas Mitzschke studied technology, knowledge and practices of agriculture, European politics, political sociology, and science and technology studies at Maastricht University (the Netherlands), the University of Essex (UK) and the Xavier Institute of Management, Bhubaneswar (India). He is currently writing his PhD thesis on the public controversies about genetically modified crops in India and Europe to understand how crop biotechnology and public involvement in this issue shape democratic political cultures in a globalized world. Mitzschke argues for the explicit inclusion of normative issues of justice, solidarity and democracy in current debates about technological risks in the global North and South.

Tom Moylan is the Glucksman Professor Emeritus in the School of Culture and Communications and an adjunct professor at the School of Architecture of the University of Limerick (SAUL). He is a cultural studies and literary scholar who studies utopianism, in the form of literary and texts and political and social practices. He is the founder of the Ralahine Centre for Utopian Studies, where he is also one of the editors of the Ralahine Utopian Studies Book Series. He has published two monographs on utopian and dystopian science fiction: *Demand the Impossible: Science Fiction and the Utopian Imagination* and *Scraps of the Untainted Sky: Science Fiction, Utopia, Dystopia.* They were breakthrough studies not only of utopian and dystopian fiction but more deeply of the critical capacity of the utopian method as a force for transformative social change. He has published numerous essays on utopia, dystopia, theology, pedagogy and political agency and he is the co-editor of several volumes, including *Not Yet: Reconsidering Ernst Bloch*, *Dark Horizons: Science Fiction and the Dystopian Imagination* and *Utopia-Method-Vision: The Use Value of Social Dreaming*. He has co-edited special issues of *Utopian Studies* on Ernst Bloch, Fredric Jameson, Irish utopias and utopia and music.

Ari Schick is a postdoctoral fellow at the Edmond J. Safra Center for Ethics at Tel Aviv University, as part of its research program on science, ethics and democracy. His research is focused on bioethics at the intersection of science, technology and culture. His recent publications have looked at speculative bioethics and science fiction, and ritual as a component of religious ethics. He completed his Doctorate in Philosophy at Michigan State University.

Barbara Segaert holds a Master's Diploma in Oriental Studies, Islamic Studies and Arab Philology from KU Leuven, Belgium, and a Master's Degree in the Social Sciences from the Open University, UK. Since 2002 she has been scientific coordinator at the University Centre Saint-Ignatius Antwerp, where she develops academic programmes on various topics of contemporary relevance to society.

Frédéric Vandermoere is Associate Professor in the Department of Sociology at the University of Antwerp. His research interests are located in the fields of science and technology studies, sustainability transitions, the sociology of risk and uncertainty, organization studies and public sociology.

Gert Verschraegen is Associate Professor of sociology in the Department of Sociology, University of Antwerp. His research interests include science and technology studies, social theory, the sociology of human rights and the sociology of Europeanization.

Thomas Völker is a postdoctoral researcher at the Joint Research Centre of the European Commission in Ispra, Italy. He holds degrees in sociology and science and technology studies from the University of Vienna. In his PhD thesis he focused on practices of producing and circulating anticipatory knowledge in transdisciplinary sustainability research and on the simultaneous rehearsal, stabilization and contestation of collectively imagined futures. His research interests include collaborative future-making in science and technology and the spatialities and temporalities of participatory research. Apart from that, he is fascinated by science fiction (SF) novels and the multiple associations of SF and emerging technoscientific fields.

Alexander Wentland is a postdoctoral researcher in the Munich Center for Technology in Society (MCTS) at the Technical University of Munich, where he works on projects concerning travelling imaginaries of innovation and contemporary technopolitics. Alexander's research is mostly focused on innovation, sustainability transitions and mobility futures. In his thesis at the Technical University of Berlin, he examined the relationship between political discourse, local assemblages and cultural imagination related to the electrification of transportation in Germany. Alexander is also affiliated with the Science Policy Research Group at the Berlin Social Science Center (WZB) and the Program on Science, Technology and Society at Harvard University.

Acknowledgements

This collection of essays emerged out of a two-day workshop at the University of Antwerp, during the spring term of 2015. This interdisciplinary workshop was titled 'Science shaping the world of tomorrow. Scientific Imagination and Development of Society' and aimed at examining the role of scientific imagination in the construction of societies. The two-day workshop brought over 30 researchers from Europe and the United States together to discuss their research projects on the role of imagination in science and society. Leading scholars in the field of science and technology studies, such as Sheila Jasanoff and Peter Galison, science fiction and utopia (Tom Moylan) and environmental sociology (Matthias Gross) introduced the four sessions, with a keynote presentation, followed by a panel of four paper presentations.

We are grateful to all who gave papers in the original workshop and to everyone who attended. We owe an immense debt of gratitude to the whole team at the University Centre Saint-Ignatius Antwerp for organizing the workshop and for making this volume possible. We are additionally very grateful to the other members of the scientific board: Arthur Cools, Raf de Bont, Jean Paul Van Bendegem, Maarten Van Dyck and Geert Vanpaemel. Special thanks are due to our authors for taking part in the project and making this volume possible. Finally, many thanks are owed to Christel Van Wonterghem (UCSIA) and to Elena Chiu and the rest of the team at Routledge for bringing this book to fruition.

Gert Verschraegen
Frédéric Vandermoere
Luc Braeckmans
Barbara Segaert
Antwerp, August 2016

1 Introduction

Shaping the future through imaginaries of science, technology and society

Gert Verschraegen and Frédéric Vandermoere

In many cultures the past is made the locus of 'a golden age', but modern society has likewise made the future into a haven for unfulfilled ideals. At least since the Enlightenment, modernity has been seen as an emancipation from a stagnant and archaic past, which explains the value attributed to thoughts, visual depictions and language trying to imagine the future towards which we travel. As the grand works of 'science fiction' testify, modern societies are fascinated by stories of how science and technology create new possibilities, yet also new risks. Although most people regard science fiction as a highly entertaining, but also partly superfluous and untrustworthy medium, this genre also points to the deep entanglement of the present with our imagination of the future. Science fiction stories bear witness to the hopes and anxieties that are created by the accelerating scientific, technological and social changes and underline the collective belief that science and technology are fundamental when it comes to steering society in the direction of a desired future.

The conviction that societies can be improved through scientific research and technological imagination is part and parcel of modern society. 'Progress' first became a key leitmotif during the seventeenth century, when reformers declared that the best way forward was to encourage science and reason in order to plan and shape the future. Historians and sociologists have shown how this understanding of the future as an open space for events that can be influenced through deliberate action only emerged with the advent of modernity (Koselleck, 2004; Luhmann, 1976). Until the beginning of the seventeenth century, perceptions of the future were characterized mainly by theological expectations of the world's immanent end. In such a religious world view, the idea of the future being a realm that could be shaped through human action and planning was impossible to conceive (Beckert, 2016, p. 28). With the emergence of the Enlightenment notion of progress, however, the horizon of the future became open and dynamic. Spurred by the rapid scientific, technological and economic changes of the time, the hope emerged that, although the future is uncertain, it can be helped into existence by relying on scientific and technological development, calculation and instruments of prognosis. This interest in envisioning and trying to shape the future has never really vanished from modern society.

In our age of rapid technological and scientific transformation, there is a renewed concern with future(s). Imagining and forecasting new futures has become an intrinsic part of our technological societies: how will we develop and plan smart cities and mobilities? How can we roll out renewable energies to tackle global warming? What new biological entities can we develop? Science and technology play a crucial role in our collective attempt to make sense of the future. Technological developments such as nanotechnology, robotics or solar energy shape how we dream and think about the future, while economic forecasts, gene tests or climate change projections help us to make images of what may possibly occur in the short- or long-term future. Our societies and organizations try to imagine what our social and technological futures will look like, yet also work upon this imagination by taking actions in organizations, drawing up plans and taking decisions about financing.

This book provides one of the first interdisciplinary assessments of how scientific and technological imaginations matter in the formation of human, ecological and societal futures. Rooted in different disciplines such as sociology, philosophy and science and technology studies, it explores what various actors such as scientists, companies or states imagine the future to be like and how they act upon that imagination. Bringing together case studies from different regions around the globe, including the electrification of German car infrastructure or the genetic modification of crops in India, it shows how sociotechnological futures are imagined and how these imaginaries enable dynamics of social change and choice, opening horizons towards alternative futures.

In this introductory chapter, we aim to set the scene for the other contributions by giving a short overview of the current discussion on collective imaginaries and future-making. We first discuss the social and intellectual context within which the technoscientific imagination has become important for future-making and engage with some existing approaches towards the idea of future-making within the social sciences. We then present the different contributions to this book and lay out the different thematic connections between them.

The social bases of imaginaries of the future

While research in the social sciences traditionally explained individual and collective action by referring to the past and the force of social reproduction, the last decennia have seen an increased awareness of how expectations of the future shape present actions. In the fields of science and technology studies, as well as in the social sciences more broadly, social, economic and technical processes are seen as being driven by collective imaginations of the future (e.g. Borup *et al.*, 2006; Brown, Rappert and Webster, 2000; Jasanoff and Kim, 2016). The decisions of individuals, groups and organizations are not solely determined by previous experiences and existing structures, but 'are shaped in equal measure by perceptions of the future' (Beckert, 2016, p. 35).

An array of scholars has emphasized the importance of imagination as a vital component of these future-oriented representations and projects. Whereas much of modern science and modern thinking has been preoccupied with constructing the world as a body of observable facts, which should be neatly distinguished from fantasy and imagination, contemporary literature has emphasized how our ideas about reality, and particularly about the future, are always bound up with the force of imagination. An ability to imagine the future is crucial in overcoming the uncertainty stemming from the openness of the future. Having a sense of how the future *can* look provides orientation in decision-making despite the unknowability of the future (Beckert, 2013, 2016).

It is important to emphasize that actors' images of how the world will look at a future point in time are collective performances. There has been a growing literature on how the capacity to imagine futures is a crucial resource in political and social life, in which imagination is no longer seen as an individual capacity but as a collective phenomenon, organized through common practices. Influential thinkers such as Benedict Anderson (1983), Arjan Appadurai (1996) and Charles Taylor (2004) paid attention to the constitutive powers of collective imagination in the emergence of the nation state, the modern moral order or the contemporary landscape of globalization. In his now classic *Imagined Communities* (1983), Anderson famously described how new social practices like reading the newspaper enabled the emergence of a sense of shared destiny and community typical for the nation state. By reading the newspaper citizens could 'imagine' themselves to be in a community of other citizens reading the paper – an imaginary community, as most members never actually met one another. In a sense, Arjun Appadurai (1996) extended the historical work of Anderson by applying the notion of imaginaries to the contemporary, globalized world. Whereas print capitalism was crucial for the rise of the nation state, we have now entered an era in which digital media and mass migration are bound to have an equally profound impact on the way in which we make sense of the world we live in. In his own way, Charles Taylor expanded the analysis of collective imaginations to address the emergence of grand patterns of moral and political thought. Taylor defined social imaginaries as:

> the ways people imagine their social existence, how they fit together with others, how things go on between them and their fellows, the expectations which are normally met, and the deeper normative notions and images which underlie these expectations.
>
> (2004, p. 23)

Imaginaries in Taylor's understanding are a set of collective assumptions and expectations that make possible and legitimize how collectivities order themselves. They are somewhat indefinite or inarticulate understandings enabling a 'wider grasp of our predicament: how we stand to each other, how we got to where we are, how we relate to other groups' (Taylor, 2004, p. 25). Imaginaries are hence not individual constructions, but pertain to the imaginations of large groups, and can thus hold large things together, such as nationhood or modern democratic space.

Imaginaries of scientific and technological futures

Recently, scholars from sociology and economics have highlighted the importance of social imaginaries in the structuring of social and economic life. In economics, authors such as George Shackle (1979) have emphasized the importance of collective imaginaries in the economy, where consumers, investors and corporations constantly have to take decisions about an uncertain future. In sociology, such scholars as Jens Beckert (2016), Niklas Luhmann (1995) and Ann Mische (2009) have paid attention to the role of expectations, anticipations and aspirations towards the future in shaping social life. Yet, as Sheila Jasanoff remarks, an 'almost inexplicable omission from most accounts of social imaginaries is a detailed investigation of modernity's two most salient forces: science and technology' (2016, p. 8). Partly in reaction to this 'systematic obscuration' (Jasanoff, 2016, p. 8) of the role of science and technology, the last decennium has seen new strands of scholarship on future imaginaries emerge from science and technology studies.

Particularly influential has been the notion of 'sociotechnical imaginaries', which points to the importance of deeply institutionalized collective visions of what a good and attainable future looks like as they are 'reflected in the design and fulfilment of nation-specific scientific and/or technological projects' (Jasanoff and Kim 2009, p. 120). By paying attention to how future expectations and public forms of reasoning shape practices and policies of science and innovation, Jasanoff and Kim have shown how culturally specific imaginations are not purely mentalist inventions but 'penetrate the very designs and practices of scientific research and technological development' (2009, p. 124). In line with recent thinking in the field of science and technology studies, they emphasize how social and political imagination is co-produced with technoscientific practices and organization.

In the field of science, technology and innovation studies, imaginaries or anticipatory visions of the future have by now become an important field of research. An array of empirical studies has shown how collective visions and expectations about future scientific or technological developments shape and influence whether and how these developments come about. Imaginaries of the future are in part 'make-believe games' (Beckert, 2016; Walton, 1990) that constitute narratives about how new technologies and innovations will change our lives and societies. When these stories are 'taken up in statements, brief stories or scenarios' (Van Lente and Rip, 1998, p. 205), they have the power to 'influence technological design, channel public expenditures, and justify the inclusion or exclusion of citizens with respect to the benefits of technological progress' (Jasanoff and Kim, 2009, p. 121). Imaginaries of the future can thus be seen as 'productive' or 'performative', in the sense that they shape the conditions of possibility for the further development of science and technology. Future science and technology are often presented as 'the "place" where solutions are realized, presences manifested, and wrongs righted' (Michael, 2000, p. 22). Promissory stories about future technologies and innovations can constitute a major resource for actors to draw on when trying to enrol other actors into networks of scientific or technological innovation (Fujimura, 2003; Van Lente and Rip, 1998). Particularly in the early stages of a technological or scientific development, promissory images of the future are constitutive in 'attracting the interest of necessary allies (various actors in innovation networks, investors, regulatory actors, users, etc.) and in defining roles and in building mutually binding obligations and agendas' (Borup *et al.*, 2006). Voicing high expectations and selling promises are common practices in new and emerging science and technologies (NEST) such as nanotechnologies, as they can mobilize (social, material and financial) resources for a protected space in which actors can collaborate to bring about what a story has anticipated (Van Lente, 1993). 'Whether imaginary and outcome actually correspond,' Jens Beckert rightly notes, 'must be answered empirically. By the same token, expectations can collapse if promissory stories cease to be convincing' (Beckert, 2016, p. 177).

Uncertainty and the politics of future-making

Making promissory stories about future scientific and technological developments credible and obtaining enduring support to channel resources into risky projects is a huge challenge. Scientific and technological imaginaries need to become materialized and converted into durable things and identities. Obviously, this process is not without friction or resistance. Competing imaginations, some more powerful than others, mostly try to establish themselves on the same terrain. Beckert (2016, p. 184) correctly notes that 'if resources for innovation are allocated based on promissory stories whose future success is uncertain, then actors will inevitably contest not just the distribution of these resources, but also the imaginaries surrounding innovations'. A strand of literature hence emphasizes how sociotechnical imaginaries are subject to political struggle and are reflecting inequalities of power and authority (e.g. Brown, Rappert and Webster, 2000; Moreira and Palladino, 2005; Wilkie and Michael, 2009). Different 'disciplines, capacities and actors compete for the right to represent near and far term developments' (Brown, Rappert and Webster, 2000, p. 5), aiming to influence particular future political and economic paths while bracketing off others.

Another strand of literature has focused rather on how actors, within the confines of existing social and institutional structures, come to terms in a practical way with the openness and inherent uncertainty of the future (e.g. Beckert, 2013, 2016; Gross, 2010). Social actors have to almost continuously make decisions with regard to investments, technological developments and future actions without knowledge about what exactly will happen in the future. How do actors generate a sense of what the future will look like and become convinced that the risks that are connected to their decisions are worth taking? A dominant approach to this question is rooted in economics and assumes that decisions involving the future must be based on a rational account of expected future payoffs and costs. Another approach, which is more associated with so-called 'future studies', tries to 'manage' the uncertainty of the future through techniques such as horizon-scanning, scenario building and future visioning. Due to the fact that there is a public demand for 'scientifically based' predictions, particularly with regard to new and growing uncertainties involved, a 'forecasting industry' came into being from the 1960s and 1970s onwards. Scenario techniques, for instance, have become increasingly popular in economic planning, political consulting or technological innovation policies. Although these forecasting techniques and calculation devices shape the actual expectations of actors as well as their sense of the means by which the future can best be approached, they have important limits. Actors may try to reduce the uncertainty entailed in the future by forecasting or calculating 'most likely' scenarios, but they will never be able to eliminate the 'fundamental uncertainty' of other actors' choices, unpredictable innovations and unforeseeable side effects. Imaginaries of the future are hence crucial in overcoming the uncertainty stemming from the inherent openness of the future, and 'provide orientation in decision-making despite the uncertainty inherent in the situation' (Beckert, 2013, p. 222).

Overview of the book

The chapters in this book show how the future is imagined, brought into being and performed in various ways by a diversity of actors, practices and discourses. The contributions focus primarily on how futures are imagined within science and technology and on how these imaginaries enable dynamics of social change and choice. How do specific scientific and technological projects imagine and negotiate new potentialities (or future damage)? How do various social actors contest and struggle over desirable futures, based on competing analyses of the current and ideal state of affairs? What predictive technologies are being used to reduce the uncertainty in the future? And how do imaginaries use the past and present in order to envision and fabricate futures?

Purposefully including different disciplines such as sociology, philosophy and the field of science, technology and society, the book develops central aspects of the current research agenda for understanding the role and position of collective imaginaries in shaping societal and human futures. This is done through a systematic focus on two important dimensions of imaginaries, which have until now received little attention.

First, most contributions aim to identify the different levels and scales at which imaginaries can exist. Central to our contemporary globalized present is the existence of global imaginary futures. We can speak of global sociotechnical imaginations when 'all or at least a significant number of people across the globe share, in the short and the long term' (Adam, 1995, p. 116) similar expectations towards the future. The most pertinent example of such a globalized future imaginary is, of course, global warming, which constitutes 'a global future for all irrespective of location, nationality and political persuasion' (Adam, 1995, p. 117). Different imaginaries of the role of science and technological innovation are part and parcel of this future.

Although future imaginaries about global warming are inherently global, they can be reinterpreted and localized within specific national or local settings. Sheila Jasanoff has famously shown how sociotechnical imaginaries are often entangled with national institutions and with a national cultural repertoire about scientific and technological progress. There are 'distinctive national visions of desirable futures driven by science and technology' (Jasanoff and Kim, 2009, p. 121), which can account for differences in how nation states engage with new technological developments such as nuclear energy, nanotechnology or genetically modified food (for an overview, see Jasanoff, 2016). Several contributions to this volume also approach the topic of imaginaries from a comparative or national angle. For example, in her historical account of the Belgian population censuses, Kaat Louckx examines the ways in which population statistics are intimately connected with the formation of the nation state.

Next to macro-level analyses of imaginaries, some contributions to this book explore the micro dimensions and local levels at which imaginaries are being constructed. For example, in his analysis of an Austrian research programme, Thomas Völker shows how a national scientific imaginary in the making is constructed from the bottom up. As multiple imaginaries are operating both within and between different contexts, this book further examines the ways in which imaginaries can be located at different levels and interact with one another. For example, in his contribution about 'the automobile nation Germany', Alexander Wentland explores recent changes in the narratives about the electrification of cars, by focusing on the particular case of Germany – yet also examining how national narratives interact with hegemonic global imaginaries about technological innovation and sustainability. The chapter by Wentland also shows how new events may reactivate older images (about the electric car), which brings us to the second dimension of imaginaries that underlies and drives the contributions to this volume.

Next to the scaled nature of imaginaries, this volume devotes explicit attention to the temporal nature of sociotechnical (and other) imaginaries. It does so by exploring how the future is being anticipated and generated in the present, but also by looking back at the projections of a future in the past. Futures have a history, but the present is not only shaped by past imaginaries of the future. Present imaginaries of the future, we now know, retroact on the present. Anticipations and visions of the future produce results now, affecting what will actually be.

The different contributions to this book draw on imaginaries that by now have become the past; they analyse how current imaginaries of the future draw on past future images or refer to earlier plans for the future. Past imaginaries, sometimes long forgotten, guided decisions in the past and in that sense shaped the path to the present in which we now live. In addition, the different contributions to this book explore whether and how collective projections of the future are being transformed into concrete realities. Andreas Mitzschke, for example, studies how the debate on genetically modified organisms is embedded within different histories of agricultural development. Tom Moylan examines the role of science fiction as both critique on modern life and utopian anticipation of an improved life.

The structure of this book is organized into three parts. In Part I, the importance of imaginaries in shaping human nature is examined. The interest in the structure of human life, as well the striving for the improvement of human abilities, is ancient. Philosophers and social theorists have always delved into the material and idealistic aspects of human nature. Yet, recent advances in genetic technologies and concomitant possibilities for human enhancement have turned our stock of knowledge and imagination of the human future upside down. Research projects on the human (epi)genome are now driven both by scientific advances and future expectations of human development. This co-evolved with the rise and institutionalization of new fields of study on the law, governance and ethics of biomedical advances, which all play their part in the struggle to shape the future.

It is against this background that the second chapter, by Ari Schick, on bioethics and the regulation of the imagined future, can be situated. Schick argues that bioethics, by locating morally problematic aspects of developing technologies in the imagined future, rather than in the present, has often served to legitimate the values that underlie ongoing research programmes instead of critically studying them. He therefore proposes a shift from the dominant regulatory mode of bioethics – which tends to ask what we ought to do with the techniques we have or will have – to a critical prophetic mode that questions, in the first place, the technologies we ought to have and why.

The topic of human enhancement is further examined in the third chapter but this time from an epigenetic angle. In 'The new biology of the social: Shaping humans' future, science and public health', Jan Baedke discusses the relation between the imagination of human enhancement and the framework of postgenomics. Baedke shows how this postgenomic future is rooted in a new entanglement of biological concepts (such as phenotypic plasticity and organism) and social concepts (like self-determination and responsibility). The new synthesis between the social and the biological is historically contextualized by building on Huxley's transhumanist view of social enhancement.

Kim Hendrickx ends the first part of the book by looking at the hidden role of imagination along the food–drug divide. In the search for what he describes as 'bodies of evidence', Hendrickx illustrates how our imagination can gain more political momentum if we take uncertainty, rather than certainty, as the starting point of our investigation.

In Part II, the importance of imaginaries in shaping techno-natures is examined. In the last decades, authors such as Bruno Latour and Donna Haraway have argued that nature, technology and society do not exist independently of each other but are intricately tangled in mutually constituted techno-natural assemblages. Questioning the artificial ontological divides between nature, society and technology, the different contributions in this part investigate how collective understandings of the future bring to life new understandings of our environment and reshape the relationship between technoscience, nature and society.

Genetic modification is perhaps one of the ultimate examples where the classic distinctions between technology, society and nature are collapsing. In the chapter by Andreas Mitzschke, the use of imaginaries is discussed using the case of Indian and European debates about GM crops. Mitzschke first argues that the concept of sociotechnical imaginary allows us to address the competing normative dimensions of the GM debate. He further points at the temporal nature of imaginaries in normative struggles. Different views on the potential of GM crops are placed within different histories of agricultural development. Imaginaries, he demonstrates, aim at shaping the present and yet they are framed by various interpretations of the past.

In the next chapter, Thomas Völker further builds on the recent literature about sociotechnical imaginaries to study how science–society relations are imagined in case of sustainability research. Reporting on a case study about the Austrian 'proVISION' research programme, Völker shows how this funding scheme can be read as an attempt to establish an alternative imaginary of science in society. Specific attention is paid to the creation process of an imaginary. Studying 'imaginaries in the making' is a perfect way, Völker argues, to understand the dynamic and contested nature of imaginaries and the mechanisms through which certain imaginations are being collectively stabilized at the expense of others.

In 'An automobile nation at the crossroads: Reimagining Germany's car society through the electrification of transportation', Alexander Wentland discusses how attempts to reimagine mobility are overshadowed by a 'history of failures' that dominates the common narrative around the electric vehicle. Wentland argues that, somewhat paradoxically, the failures in production and marketing have also created windows of opportunity. For example, the 'car of the future' out there challenges existing business models, user cultures and routines. In the last decennium, Wentland argues, a range of competing electrification futures has emerged, each articulating what a desirable future of mobility should look like. Every future scenario, however, requires the consolidation of a certain set of actors, artefacts and ideologies. Wentland claims that the continuous enactment of imagined futures is crucial to keep the project of the electric car alive.

Part III looks at the importance of imaginaries in shaping societies. Ever since the arrival of scientific modernity, the capacity to imagine, anticipate and give visibility to the future has played a prominent part in the formation of communities, cities or nation states. Drawing on specialized techniques for counting, mapping or visualizing, humans have tried to improve their abilities to model and predict the future. The contributions in this last part of the book investigate how such attempts to gain insight into the future are always entangled with collective imaginaries of the social order.

In 'Parameters of nation-ness and citizenship in Belgium (1846–1947)', Kaat Louckx examines the relation between statistics and the nation state. The formation of nation states, she argues, partially builds on the use of population statistics. Louckx describes the 'scientific' representations of modes of belonging to the nation state as produced in the Belgian population censuses. It is argued that the ways in which inclusion in and exclusion from the nation state are imagined is reflected by the citizenship categories in the censuses. By analysing the statisticians' shifting interests, techniques and classification schemes, Louckx thus reveals the premises and principles underlying contemporary ideas and ideals of modern government.

Tom Moylan discusses two major works of science fiction (Shelley's *Frankenstein* and Piercy's *Woman on the Edge of Time*) in order to illustrate how this genre draws from and speaks back to science and technology. Through the combination of perspectives from literary and utopian studies he eloquently describes how the sciences and scientific imagination come to shape social imagination and vice versa. More particularly, it is shown how science fiction – as an imaginative approach that refuses to regard the present as static – can produce both a critique of the present and a utopian anticipation of an improved life for humans.

In the final chapter of this book, titled 'Shaping new horizons: Proactionary attitudes, precautionary principles and the experimentalities of science in society', Matthias Gross notes that the use of experimental strategies has become common among practitioners and engineers. In contrast to precautionary measures, he makes a plea for a 'proactionary principle'. Using results from two case studies – one about contaminated industrial sites, a second about the search for geothermal heat – Gross shows how experimentality is becoming a defining feature of modern society. Under the condition that 'adventurous innovations' become socially just, ignorance can serve as a productive resource.

Obviously no single collection of essays can capture the full richness and nuances of a rapidly evolving debate such as the current discussion on scientific and technological futures, even more so because the scope of this collection is not limited to a single country, policy area or specific time. Indeed, as the chapters in this volume cross borders, the authors drive home the need to embrace the complexity of our collective imaginations of the future. The future entails more than calculated risks and profits – it includes the full force of the creativity and imagination of science and technology.

References

Adam, B., 1995. *Timewatch. The social analysis of time*. Cambridge: Polity.

Anderson, B., 1983. *Imagined communities*. London: Verso.

Appadurai, A., 1996. *Modernity at large. Cultural dimensions of globalization*. Minneapolis, MN: University of Minnesota Press.

Beckert, J., 2013. Imagined futures: fictional expectations in the economy. *Theory and Society*, 42, pp. 219–240.

Beckert, J., 2016. *Imagined futures. Fictional expectations and capitalist dynamics*. Cambridge, MA: Harvard University Press.

Borup, M., Brown, N., Konrad, K. and Van Lente, H., 2006. The sociology of expectations in science and technology. *Technology Analysis & Strategic Management*, 18(3–4), pp. 285–298.

Brown, N., Rappert, B. and Webster, A., 2000. *Contested futures: a sociology of prospective techno-science*. Burlington, VT: Ashgate.

Fujimura, J., 2003. Future imaginaries: genome scientists as sociocultural entrepreneurs. In: A.H. Goodman, D. Heath and M.S. Lindee, eds. 2003. *Genetic nature/culture: anthropology and science between the two-culture divide*. Berkeley: University of California Press. pp. 176–199.

Gross, M., 2010. *Ignorance and surprise: science, society, and ecological design*. Cambridge, MA: MIT Press.

Jasanoff, S., 2016. Future imperfect: science, technology and the imaginations of modernity. In: S. Jasanoff and S.-H. Kim, eds. 2016. *Dreamscapes of modernity. Sociotechnical imaginaries and the fabrication of power*. Chicago, IL: University of Chicago Press. pp. 1–33.

Jasanoff, S. and Kim, S.-H., 2009. Containing the atom: sociotechnical imaginaries and nuclear regulation in the U.S. and South Korea. *Minerva*, 47(2), pp. 119–146.

Jasanoff, S. and Kim, S.-H. eds. 2016. *Dreamscapes of modernity. sociotechnical imaginaries and the fabrication of power*. Chicago, IL: University of Chicago Press.

Koselleck, R., 2004. *Futures past: on the semantics of historical time*. New York, NY: Columbia University Press.

Luhmann, N., 1976. The future cannot begin: temporal structures in modern society. *Social Research*, 43, pp. 130–152.

Luhmann, N., 1995. *Social systems*. Stanford, CA: Stanford University Press.

Michael, M., 2000. Futures of the present: from performativity to prehension. In: N. Brown, B. Rappert and A. Webster, eds. 2000. *Contested futures: a sociology of prospective techno-science*. Aldershot: Ashgate. pp. 21–39.

Mische, A., 2009. Projects and possibilities. Researching futures in action. *Sociological Forum*, 24, pp. 694–704.

Moreira, T. and Palladino, P., 2005. Between truth and hope: on Parkinson's disease, neurotransplantation and the production of the 'self'. *History of the Human Sciences*, 18(3), pp. 55–82.

Shackle, G.L.S., 1979. *Imagination and the nature of choice*. Edinburgh: Edinburgh University Press.

Taylor, C., 2004. *Modern social imaginaries*. Durham, NC, and London: Duke University Press.

Van Lente, H., 1993. *Promising technology: the dynamics of expectations in technological development*. Delft: Eburon.

Van Lente, H. and Rip, A., 1998. Expectations in technological development: an example of prospective structures to be filled in by agency. In: C. Disco and B. van der Meulen, eds., 1998. *Getting new technologies together: studies in making sociotechnical order*. Berlin: De Gruyter. pp. 203–229.

Walton, K.L., 1990. *Mimesis as make believe: on the foundations of the representational arts*. Cambridge, MA: Harvard University Press.

Wilkie, A. and Michael, M., 2009. Expectation and mobilisation. Enacting future users. *Science, Technology, & Human Values*, 34(4), pp. 502–522.

Part I

Shaping human nature

2 Biothics and the legitimation/ regulation of the imagined future

Ari Schick

Introduction

Speculative bioethics and human enhancement

Although bioethics is primarily organized around applied issues in biomedicine (principally research ethics, clinical ethics and the use and governance of biotechnology), ever since the field became established in the United States roughly 50 years ago it has included a speculative side drawn to technological futures. Initially, speculative bioethics was tied to specific technological possibilities such as artificial methods of reproduction, human cloning and genetic engineering. However, over the past two decades a unifying rubric of human enhancement has emerged, linking speculative discussions that range across a variety of technological possibilities such as pharmaceuticals, human–machine integration, genetic engineering and various possibilities subsumed under the heading of nanotechnology.[1]

Human enhancement has now become a prominent topic of bioethics discourse, covering not only possible future technologies but also a selection of existing technologies and practices, such as drugs that can improve mood or boost performance on cognitive or physical tasks. Even so, technological imaginaries are essential to how the discourse is structured, for the true concern and motivation for addressing human enhancement in a manner that extends well into the possible future lies in the idea that the present can be understood as a prelude to futures in which far more powerful enhancement technologies will exist.[2] In these scenarios, enhancement technologies have become available that can significantly extend human capabilities or modify behavioural tendencies in ways that may change the nature of intimate relationships, affect individuals' self-understanding or disrupt once-stable configurations of the social world. It is discussions of these kinds of possible technologies – characterized as disruptive or transformative and marked by high levels of both technical and social uncertainty – that I have in mind by the use of the term 'speculative bioethics'.

As in all discussions of bioethics, several clarifying points are necessary. First, 'bioethics' is a term that ranges over a rather diverse set of people, institutions, practices and areas of inquiry, and therefore speaking generally about bioethics introduces its own problems. This chapter addresses an aspect of 'public bioethics' in which issues at the intersection of social values, governance and public policy are discussed. Second, although a noticeable portion of the literature on human enhancement is now written outside the field of bioethics proper (see n. 3), because much of it is in dialogue with existing bioethics literature, I include it under the label of speculative bioethics.

Questioning speculative bioethics

Since these future technologies may vastly reshape the self and the world as we know it, some sort of proactive ethical intervention appears warranted – especially if the alternative is to allow biotechnology to progress without the benefit of ethical foresight. As reasonable as this seems, the logic of the speculative bioethics enterprise is unclear. Who establishes the content of these futures and by what method? Is the intent to provide ethical guidance for the development of technologies or to govern how they are used once they arrive? What happens when we link ethical questions regarding existing technologies to possible future technological developments? What can bioethics contribute to an already-crowded arena of expert technological future-gazing? Although not all of these points can be addressed here in detail, they motivate the present discussion.

Despite the salience of such questions, among the varied practices, institutions and discourses that constitute bioethics today, speculative bioethics may well be the least explicitly analysed – having been mostly overlooked as a discrete aspect of the bioethics ecosystem both by social scientists, who have studied, critiqued and (one can hope) promoted greater reflexivity within bioethics, and by those ethicists who have been alert to issues of purpose and methodology within their field.[3] Although there are scattered critiques of speculative bioethics from within, systematic discussions have mostly been lacking.[4] Similarly, there is little outside bioethics that directly addresses why this applied field so often delves into highly speculative futures or seeks to evaluate the consequences of exploring these kinds of technological imaginaries.[5]

Aims and structure of the chapter

This chapter examines speculative bioethics by integrating a critical perspective rooted within bioethics, together with insights from science and technology studies (STS). The former builds on bioethics' own tradition of reflection on its history, goals and methods as a publicly engaged and ultimately practical form of public discourse; the latter attends to how the production of technoscience is intertwined with shared technological expectations and cultural imaginaries that are contested and legitimated via various expert and lay discourses and within aspects of popular culture. This scholarship is essential to a critical examination of speculative bioethics, for it allows us to situate bioethical discussions concerned with technological futures among various other overlapping discourses that are similarly motivated and structured by emerging technoscience, and which may in turn affect how agency is configured across a range of domains (Brown, Rappert and Webster, 2000; Nordmann, 2010).

The approach I take is particularly informed by work that explores how expectations for future technoscientific developments capture public attention, help to marshal support for scientific research and enable coordination between diverse sets of actors, but are communicated using varying levels of certainty depending on the context (Borup *et al.*, 2006; Brown and Michael, 2003). I also draw on recent STS literature that has begun to examine the broader significance of shared technologically imagined futures (sociotechnical imaginaries) within contemporary societies as both encoding visions of the social good and mediating collective agency (Jasanoff and Kim, 2015).[6] Sociotechnical imaginaries should therefore be of as much concern to ethicists as social scientists and historians.

In what follows, I first briefly trace the development of speculative bioethics in relation to different modes of bioethics discourse, highlighting the key changes that have taken place over the past three decades. I then offer an analysis of the current state of speculative discourse, and argue that the dominant discursive mode of engagement, here termed 'anticipatory ethics', is problematic and of dubious value. Finally, I argue that bioethics' engagement with speculative futures can be justified and constructive in theory, but would need to combine resources already found within bioethics with others that it has yet to develop. The model of speculative bioethics I offer is better attuned to the unpredictability of scientific progress and sensitive to the nature and function of sociotechnical imaginaries.

A critical overview of speculative bioethics

Technological imaginaries and the establishment of bioethics

The 1960s and 1970s – the period in which bioethics coalesced into a recognizable field in the United States – were rife with speculation and anxiety regarding the unprecedented degrees of control over human reproduction and heredity that might be achieved through the rapidly advancing fields of embryology and genetics.[7] Although visionary scientists and other intellectual elites had been discussing in earnest the possibilities of *in vitro* fertilization, artificial wombs, cloning and directing human evolution since the 1920s (McLaren, 2012; Squier, 1994; Tirosh-Samuelson, 2012), the new molecular genetics was seen as having the potential to bring these visions to fruition much sooner than had ever been expected. Non-scientists, including sociologists and historians (Etzioni, 1973; Goodfield, 1977), popular writers (Rosenfeld, 1969; Toffler, 1970), and religious ethicists (Fletcher, 1974; Ramsey, 1970a), soon joined this newly urgent conversation. These religious ethicists did not write solely in a speculative vein but contributed to a range of applied ethical issues in biomedicine that would come to define bioethics, such as end of life care, human subject research and organ transplantation (Ramsey, 1970b).

Bioethics came to prominence in the 1970s as the public became wary of leaving these issues solely to the discretion of scientists and clinicians. Yet, as it sought to establish research centres and secure funding, bioethics did not position itself as antagonistic to science, but as a practical endeavour that could collaborate with, and be useful to, biomedicine. Daniel Callahan, who in the late 1960s cofounded the Hastings Center, the first bioethics think tank, made this clear in his programmatic vision for the field:

> My contention is that the discipline of bioethics should be so designed, and its practitioners so trained, that it will directly – at whatever cost to disciplinary elegance – serve those physicians and biologists whose position demands that they make the practical decisions.
>
> (Callahan, 1973, p. 73)

Because research in the life sciences was now attracting significant public scrutiny, the needs of biologists went beyond making practical decisions, and included communicating to the public that they did not consider themselves to be the sole arbiters of humanity's future. Bioethics' alignment with biomedicine thus carried over into the realm of the technoscientific imagination, as is evident in a 1971 'Ethical Issues in Human Genetics' conference, jointly organized by the Hastings Center (then called the Institute of Society, Ethics and the Life Sciences) and the Fogarty International Center of the National Institutes of Health (Hilton *et al.*, 1973). This may be regarded as among the first proper bioethics conferences, in that many of the organizers and participants were already associated with the field that would soon be widely known as 'bioethics'.

The conference was notable for a spirit of collaboration between scientists, doctors, ethicists and policy experts, and by the fact that it was motivated by a concern with both newly emerging and potential future genetic technologies.[8] The preface to the conference's published proceedings, written by Callahan and biochemist Peter Condliffe of the NIH, evinces a tension between a desire to explore the anticipated technological futures that were causing public concern, and a belief that the conference ought to focus on the practical questions that were arising with newly developed technologies:

> To some of us who participated in this present meeting the debate seemed to be over this basic issue. Ought medicine through applied human genetics, to take part in [directing human evolution] by altering either the probability of change or by intervening in the selection process through negative means such as abortion, or, in the future, by directly altering the chemical structure of DNA?
>
> As Monod points out, we now understand biology sufficiently to construct a coherent theory about it.[9] This enables us to predict the probable lines of technological development in biology and medicine even though much experimental detail must be filled in. We are indeed in a situation analogous to that of the physicists following the discovery of nuclear fission by

> Hahn and Meitner, after which it was certain that the atomic bomb could be constructed if sufficient effort was made. Nirenberg has pointed out the inevitability of technological developments such as cloning, construction of synthetic genes, and other phenomena. Much of this future development is discussed in these proceedings by Sinsheimer.[10]
>
> (Condliffe and Callahan, 1973, p. x)

Using the rhetoric of technological inevitability and foreseeability, the imagined future of genetic technologies was presented as a key issue for ethicists and scientists to examine long before such capabilities are realized. Yet, ultimately, this imaginary, though addressed in several sections of the conference, was not to be the central focus.

> The conference program committee … while cognizant of these future developments, felt that examination of immediate ethical issues presented to the genetic counselor in his practice of medicine was a useful tool for getting at the existing and future predicament. Precedents set now will probably form the basis for future legislation and for legal disposition of future cases arising from the more esoteric forms of genetic intervention.
>
> (Condliffe and Callahan, 1973, pp. x–xi)

This set a course that professional bioethics would follow for roughly the next two decades. The prospect of genetic engineering – said to rival nuclear fission in its potential to affect the future of the human race – though it raised still-unresolved questions, was moved to the periphery, while emerging practical issues in genetic screening and testing were foregrounded (Jonsen, 1998, p. 180). Later, as DNA technologies and genetic therapies began to appear as nearing possibilities, a presidential bioethics advisory commission likewise stressed the importance of ensuring that clinical interventions (at the somatic level) would proceed safely with due oversight, while downplaying the necessity of proactively delving into more distant, highly controversial possibilities such as directly altering the human germline (President's Commission, 1982).

From prophetic to regulatory bioethics

The deflationary stance that bioethics adopted toward technological futures might appear to be a simple consequence of the decision to prioritize real-world issues over still-speculative ones. Yet there was a more significant cause: an underlying shift in the dominant mode of bioethics discourse that affected how technological futures could be discussed.

Broadly speaking, bioethics includes two discursive modes, described by Callahan as:

> two powerful currents in bioethics, not necessarily incompatible but surely moving in different directions. One of them turned its attention to individual

> rights and choice, with which the analytically trained philosophers and lawyers were most comfortable, and the other to the social and cultural meaning of the biomedical developments, which profited from a religious and social-science presence as well as nonanalytic philosophical approaches.
>
> (Callahan, 1999, p. 280)

Following the terminology used by Callahan (1990, 1993) and others, I refer to these as the 'regulatory' and 'prophetic' modes of bioethics, respectively.[11] Although the two modes of bioethics are 'not necessarily incompatible', in practice they come into conflict in the form of disagreements over what kinds of values and value terms are appropriate for policy deliberations within the public sphere (Moreno, 2005).

In the regulatory mode, bioethics typically functions in formal and informal capacities as a form of oversight over biomedicine. Ethicists focus on the governance of current or near-term practices, programmes and technologies in clinical medicine, biomedical research and health policy. They may formulate guidelines, point out problems and suggest reforms, but their purpose is generally not to rethink practices from the ground up or to question the wisdom of the whole enterprise. Discussions commonly proceed on the basis of rights-based moral-political concepts with widespread currency in liberal societies, with a concomitant tilt toward liberal individualism. Harms and benefits, autonomy and justice and the myriad ways these contend with each other in varying circumstances form the core of regulatory bioethics. In terms of justification, regulatory bioethics usually eschews comprehensive moral theories and proceeds on a consensus model, either consolidating what are described as widely accepted norms into mid-level ethical principles (Beauchamp and Childress, 2008) or working towards consensus in practice (Moreno, 1995). A schematic consensus position might, for example, balance a concern for risks against a perceived need to allow research to progress in the hope of producing future benefits, or place limits on individual autonomy in order to prevent harm or promote justice.

Prophetic bioethics (named after the biblical prophet's role as a social critic rather than oracle) is rather different.[12] Although it touches on many of the same issues within biomedicine, it seeks to interrogate the social and cultural meaning of biomedical practices and technologies, such as the effects of increased control over biological processes on social and political life, and on the human condition more generally. It tends to be anthropological and reflective and can be highly critical of the status quo – asking whether present developments within biomedicine and the biosocial configurations that they give rise to promote or distort particular conceptions of human flourishing and the social good. By transposing these kinds of questions into speculative discussions, theologians, ethicists and visionary scientists were able to debate the ends that science and medicine should be working towards.

Prophetic bioethics of this kind, as sociologist John Evans notes (2002), is 'substantively rational', concerned with establishing what constitutes legitimate or illegitimate ends. However, regulatory bioethics is 'formally rational' and

Table 2.1 Regulatory and prophetic bioethics[1]

Regulatory bioethics	*Prophetic bioethics*
1 General perspective: local	1 General perspective: global
1.1 Examines issues in biomedicine as they arise due to the need for practical guidance.	1.1 Explores the social and cultural meaning of various aspects of biomedicine.
1.2 Explores issues narrowly; structural issues are often not treated as relevant.	1.2 Particular questions are often signifiers of larger social issues or raise questions of society's goals and values.
2 Purpose: governance and oversight	2 Purpose: moral reflection and social critique
2.1 General acceptance of biomedical practices and assumptions regarding their positive contribution to human wellbeing.	2.1 Often questions the goals and practices of biomedicine and their impact on society.
2.2 Views itself as neutral regarding particularistic cultural and religious values and regards them as outside the bounds of public policy deliberations.	2.2 Concerned with the deep connection between values and culture.
2.3 Aims at practical agreement.	2.3 Aims to advance a particular moral viewpoint or to encourage shared moral understanding.
3 Form of moral discourse: rights-based	3 Form of moral discourse: various philosophical and religious traditions
3.1 Use of mid-level principles; shuns 'vague' value terms.	3.1 Use of 'thick' value terms.
3.2 Individual autonomy as primary value (each person chooses their own ends) tempered by demands of justice.	3.2 Substantive views on the nature of the good and the proper ends society ought to pursue.
3.3 Social good is what a society values at a given time.	3.3 Social good is deeply rooted in certain configurations of social life and the human condition.
4 Can see prophetic discourse as unfit for public policy discussions; an attempt to impose parochial values in a pluralistic society.	4 Can sees regulatory discourse as reductive, individualistic and unable to articulate notions of the common good; an attempt to strip the public square of shared values.
5 Application to future technologies: anticipatory	5 Application to future technologies: contemporary or visionary
5.1 To prepare for the eventual arrival and implementation of a future technology.	5.1 As another point of reflection on society and culture.
	5.2 As representing ends that society ought/not pursue through technology.

Notes

1 Although it is not always obvious, prophetic and regulatory bioethics should not be conflated with conservative and liberal viewpoints, respectively. There are otherwise conservative authors who see regulatory approaches as necessary in a pluralistic society and liberal authors who advance a prophetic view of collective and individual flourishing based on enlightenment conceptions of human progress.

thus limited to assessing a given research programme, medical intervention, or biotechnology as means to ends that are already accepted (Evans, 2002). As bioethics became institutionalized as a secular arbiter of public concerns regarding biomedicine, a way of developing policy guidance for advancing biomedical research, and a language of negotiating between the rights of patients and the responsibilities of healthcare providers, it became a predominantly regulatory discourse (Callahan, 1990, 1993, Evans, 2002, 2012). This marginalized engagement with the speculative sociotechnical imaginaries taken to represent the ends that society might pursue via genetic technologies and the underlying question of whether we ought to refrain from intervening in fundamental biological processes that were previously 'in God's hands', or actively pursue futures of increasing control over reproduction and heredity' (Evans, 2002).

Anticipatory ethics and the turn to human enhancement

Technological futures began to feature more centrally within bioethics from the 1990s, though it was not because prophetic discourse was renascent. Rather, the perceived distance between the present and future began to condense, and with it, bioethicists' views of which technologies warranted regulatory scrutiny began to shift ahead in time (Schick, 2016). The initiation of the Human Genome Project (HGP) and its accompanying Ethical, Legal and Social Implications (ELSI) research programme was unquestionably significant in this regard.[13] The HGP remains the largest coordinated strategic initiative in the life sciences, working towards the specific goal of sequencing a complete human genome. The bioethics initiative that it supported was similarly goal-oriented. ELSI's purpose, as expressed by Eric Juengst, its first programme director, was to 'anticipate the social consequences of the project's research and to develop policies to guide the use of the knowledge it produces' (Juengst, 1991b, p. 71).

To this end, ELSI provided an unprecedented amount of funding for forward-looking bioethics and policy analysis on genetic technologies (a subfield soon christened 'genethics'),[14] initiating a form of bioethics that was both practical and future-oriented: in effect, speculative bioethics in a regulatory mode. This new 'anticipatory ethics' viewed expected technologies and technological scenarios as futures already in the making, fully grounded in ongoing research programmes and coming ever closer to being realized. Whereas the imaginaries discussed in the earlier prophetic bioethics tended to be contested and open to utopic or dystopic interpretations, anticipatory bioethics projected regulatory interventions into technoscientific futures, envisioning the coming of technologies that produced benefits and few harms through the proactive development and implementation of ethical guidelines and oversight mechanisms.

As ELSI was getting underway, Juengst similarly advocated extending the reach of regulatory bioethics to include future germline genetic interventions. He argued that prophetic and regulatory bioethics were not parallel discourses but successive stages, and that the time had come for discussions of germline therapies to be transitioned beyond the early prophetic (what he terms 'romantic') form of bioethical discourse that could not be easily translated into 'practical policies for scientific research beyond the cautionary moratoria it has already inspired', to a 'precise' (i.e. regulatory) stage of analysis (1991a, p. 589). Even though it was too soon to attempt germline therapies in human subjects, such an eventuality was thought to be nearing, and 'as the technical potential for human germline interventions materializes, the need for such a translation becomes imminent and urgent' (Juengst, 1991a, p. 589).

Among the existing objections to germline interventions that Juengst identified as falling under a 'precise' mode of bioethics was the potential for therapeutic uses to give way to enhancement applications (whereas earlier prophetic discussions had been more concerned with the acceptability of altering genes, no matter the intent), and, since expanding the scope of regulatory bioethics to germline interventions was itself increasingly becoming an imperative (Chapman and Frankel, 2003), genetic enhancement likewise began to feature as an important topic. Enhancement was also beginning to encapsulate the broader issue of the proper goals of medicine (Kramer, 1993). Erik Parens, an ethicist on the staff of the Hastings Center, noted that between 1993 and 1997 the idea that human enhancement was too speculative to take seriously had given way to its acceptance as a wide-ranging subject for ethicists to proactively address (1998, p. S2).[15]

Towards the close of the century, as the first draft sequence of the human genome neared completion, conferences and publications devoted to human enhancement began to proliferate. Among these was a seminal book on the ethics of reproductive genetic technologies co-authored by four prominent bioethicists (Buchanan *et al.*, 2000).[16] Decrying the essentialism that had driven a decade of 'gene mania', but nonetheless asserting that 'humankind's future abilities to rewrite our genetic code are apparently limitless' (2000, p. 56), they devoted significant attention to the possibility of genetic enhancement and explicitly argued that bioethics could productively delve into speculative 'science fiction' scenarios using its existing methods.

To some degree, the sense of urgency that spurred regulatory bioethics into speculative territory is but another instance of the chronic tendency to underestimate the difficulty of moving from theory to application in technoscience – not simply due to unforeseen technical hurdles, but because the scientific knowledge base itself can develop in unexpected ways (Brown, 2003). In this case, not only was it technically difficult to translate genomic knowledge and the technics of molecular biology into clinical interventions, the HGP itself revealed underlying complexity in gene regulation and expression that had not been accounted for. However, subjacent to the normal overoptimistic assessment was a uniquely hyperbolic set of expectations fuelled by an extraordinary cultural fascination with the gene, a degree of interest in eugenics not seen since before the Second World War, and a belief that the future of medicine writ large was genetic medicine (Nelkin and Lindee, 2004). These all became bound up with the HGP, whose scale was unprecedented for the basic life sciences. The tangible progress of the HGP towards its defining goal continually closed the distance between the present and a future epoch that the completion of the HGP would launch, one in which genetic engineering would finally be possible (Hood, 2000). Because 'risk and opportunity are the flip sides of hyperbolic expectations, inflating one another in equal measure' (Brown, 2003, p. 4), these extraordinary expectations produced a correspondingly high-risk profile (largely in terms of negative social consequences). Even as the HGP attempted to manage the perception of its risks through the ELSI programme, the more speculative aspects of genethics research only reinforced the idea that genetic technologies were inevitable, rapidly nearing and likely to have far-reaching consequences for society.

In the years following the completion of the HGP, speculative bioethics has been marked by two interrelated developments. The first is an intensified aura of inescapability and urgency surrounding coming enhancement technologies. We find books on 'inevitable' genetic enhancements and the coming of the post-human (Fukuyama, 2002; Stock, 2002), reports on human enhancement nearing through the convergence of nano-, bio-, neuro- and other technologies (Roco and Bainbridge, 2003), followed by warnings that bioethics is lagging dangerously behind (Mnyusiwalla, Daar and Singer, 2003), and the rise of a more self-consciously speculative 'nanoethics' discourse heavily invested in enhancement (Lin and Allhoff, 2006). More than 30 years after Callahan and Condliffe described the inevitability of moving from *episteme* to *techne* in genetics as akin to an inexorable progression from nuclear fission to nuclear bombs, the analogy reappears with new exigency in the introduction to an anthology on the 'future of the human condition': 'The effects of genetic enhancement like the consequences of atomic fission, will last far into the future and will not be limited to localities or even large regions. Dealing with this sobering fact has recently taken on a new sense of urgency' (Baillie and Casey, 2004, p. 3).

Very much tied in to this first phenomenon was the return of a visionary style of discourse on the technological future that had mostly disappeared after the 1970s. This emerged from a somewhat improbable confluence of factors that included a newly assurgent politically active cohort of 'bio-conservative' ethicists often writing in a prophetic mode (Anderlik, 2005; Moreno, 2005), and the mainstreaming of transhumanist thought, which brought technovisionary conceptions of human progress into bioethics (Agar, 2007).[17] To some, this was a welcome opportunity to revisit open questions bypassed by regulatory discussions (Khushf, 2007), but the intense focus on technologically transformed futures seems to have further collapsed the distance between the present and the future and elided the distinction between fictional extrapolation and prediction (Clayton, 2013). In the introduction to a themed issue of the *Journal of Medicine and Philosophy* devoted to transhumanism, we find the escalating rhetoric of imminent change presented in the matter-of-fact style of an editorial voice: 'The development of emerging biotechnologies is on the verge of redesigning the boundaries of human existence' (Jotterand, 2010, p. 618).

Absent these developments, it is possible that once expectations for genetic engineering had soured, speculative bioethics would have receded or become invested in a completely different technological imaginary. Instead, even as the promises of genetic engineering went mostly unfulfilled, there was an almost seamless transition from enhancement qua genetic engineering to human enhancement more generally.[18] The current human enhancement discourse subsumes discussions of enhancement from genethics, neuroethics and nanoethics without being tethered to any one particular (potentially disappointing) technoscience. An elusive but ever nearing imaginary, in which various technologies can be used to significantly enhance people physically, cognitively, emotionally or even morally, is now the substrate upon which much of speculative bioethics proceeds. The range of evidence that can be adduced to show that the enhancement future is impending, or already here, and demanding our attention is now drawn from an increasing number of diverse sites where technoscience is produced and interpreted.

Borrowing the hedging qualifiers that appear so often in the literature on human enhancement, I would contend that human enhancement has *potentially* entered a phase in which it exists as a self-sustaining bioethics discourse. It *may* therefore be immune to the changing fortunes of any bio, nano or neuro research programme. Regardless, at least for now, disillusionment with speculative futures within bioethics seems to have been forestalled.

A critical analysis of speculative bioethics

As I discussed, following the creation of the ELSI programme, the dominant form of speculative bioethics has been anticipatory; that is, regulatory bioethics has been applied to expected future biotechnologies. Although the justification for anticipatory ethics is not often made explicit, the implied reasoning behind it is appealing: by anticipating future technological developments and their social ramifications, we can proactively establish governing principles and regulatory guidelines so that the coming technologies do not exceed the control of our considered moral judgements and produce undesirable social effects. Even if we cannot foresee all the issues in advance, we can see enough of what these technologies will be capable of to develop approaches that will be relevant when these futures arrive. In any case, what is the harm in trying?

In sum, anticipatory ethics claims to be useful by preparing ethical guidance in advance of the arrival of future biotechnologies and to have few significant downsides. In this section, I argue that the claim for future usefulness is dubious, that anticipatory ethics can in fact subvert ethical discourse by legitimating rather than interrogating technological futures, and that anticipatory ethics structures moral discourse in a way that bypasses agency in the present.

The limits of anticipatory ethics

At the heart of anticipatory ethics is a conjunction of foresight and control aligned with the strategic orientation of technoscience. Once a research programme is viewed as pursuing a specific goal, such as the ability to alter the human genome, this generates an imagined future in which that goal has been achieved and the technology is safe and effective. From there, we can imagine likely uses for the technology and their social consequences, then assess what applications are morally acceptable and which are not. The next step is to translate these assessments into guidelines that can be implemented when the technology arrives so that it is controlled from the point at which it is first implemented.

This project, though seemingly reasonable, runs into the problem at the heart of classic technology forecasting and assessment: even if we could muster the necessary degree of political will to enact various regulations in anticipation of a powerful future technology, it would still be impossible to exercise proactive control over the technological future because we cannot accurately predict the course of a truly revolutionary technology's development, let alone its social ramifications. We therefore cannot begin to control the social effects of a transformative or disruptive technology before they begin to occur, at which point the degree of control that can be exercised is far more limited (Collingridge, 1980; Nordmann, 2010).

If outright control is not possible, then perhaps the purpose of analysing speculative futures through the lens of regulatory bioethics is more modest. Engaging in anticipatory ethics prepares us or our successors to make key decisions that will arise in the future by crafting appropriate regulatory models and formulating suitable principles, or by calibrating ourselves into a state of reflective equilibrium that is attuned to the exigencies of that future.[19] Nicholas Agar, a bioethicist with fairly mainstream views on enhancement, makes this strategy explicit:

> It is better to have principles covering situations that turn out to be impossible than to have no principles for situations in which we suddenly find ourselves. Acquiring moral insurance against the many different futures that enhancement technologies might make requires that we think beyond the limits of current science. We need principles for situations that might never eventuate, but whose possibility can not be ruled out given our current state of knowledge.
>
> (Agar, 2004, p. 34)

The necessity of anticipatory ethics is apparently grounded in the notion that we might 'suddenly' find ourselves with new technologies that appear fully formed, at which point we would be severely wanting for principles up to the task of regulating them.

There appears to be little consideration of the possibility that approaches formulated well in advance could do no better – and would likely fare worse – compared to those developed incrementally as a technology is developed and realized. In order for anticipatory ethics to function as the 'moral insurance' that Agar wants, the anticipated technologies would not only need to develop roughly as predicted. In addition, we would need to: (i) achieve a sufficient grasp of the future social context of a not-yet-extant technology to offer useful ethical analysis; (ii) have justifiable confidence that future persons, despite living in a very different technological milieu, will have values and beliefs sufficiently similar to our own to make our judgements relevant for them; and (iii) have good reason to think that an engaged and responsive incrementalist bioethics would not function at least as effectively.

The idea of moral insurance actually seems to provide only a partial rationale for anticipatory ethics. It is supplemented by a worry that public discussions of genuinely novel and potentially destabilizing new technologies can be politically contentious and divisive. A page earlier, Agar offers this counterfactual:

> Suppose that prior to 1997 [the year that Dolly the cloned sheep made headlines] there had been a wide debate about human cloning, with contributions from representatives of many different moral traditions. We might now have reached some kind of moral consensus about the various uses to which the technology should be put.
>
> (Agar, 2004, p. 33)

Anticipatory ethics is therefore not only meant to save us from the potential ill effects of technologies themselves, but from heated, misinformed, and irrational public debates that seem to spring up whenever science raises the possibility of altering or circumventing aspects of human biological life so essential to the human condition as we know it that they have seldom been thought of as merely contingent. The hope, apparently, is that if a sufficient cross-section of ethicists would talk among themselves in advance and present their findings to the public, they could clarify all the relevant issues such that when the technology arrives we will have already agreed on how to use it – essentially what Juengst had in mind for germline genetic engineering more than a decade earlier.

This, too, overestimates what an anticipatory discourse could accomplish. The prospects for establishing a broad consensus regarding the uses of a deeply controversial possible future technology while the technology exists primarily as an abstraction seem remote. Those who oppose the technology don't want it at all (a position that is not intrinsically unrealistic when the technology does not yet exist) and are bypassed by discussions of sanctioned and prohibited uses. Ethicists so inclined may debate the details of appropriate applications, but in the absence of a practical issue at hand there is no incentive to establish a consensus. Meanwhile, since the public is mostly unaware of what is being developed in research labs, regulatory bodies are under no pressure to devote resources to developing guidelines for possible future technologies and legislators have no reason to invest political capital to do so either.[20]

Anticipatory bioethics as a discourse of legitimation

It might seem unremarkable that some ethicists have an exaggerated understanding of the vanguard role that the field is capable of fulfilling, but the underlying question is why bioethicists are telling – sometimes explicitly and more often implicitly – an improbable and self-contradictory story of technologies that are at once largely foreseeable but that absent a forward-looking bioethics discourse would appear so abruptly that we would find ourselves lacking adequate moral resources to deal with them. At one level, this image of the future seeks to legitimate bioethics itself (cf. Hedgecoe, 2010); significant ethical concerns will arise, but, thanks to work done by bioethicists today, the issues will be addressed calmly, systematically and without political rancour. Yet this explanation is lacking: why does bioethics seek this kind of legitimacy when it already has a degree of institutionalized recognition and public standing enjoyed by few similar fields?

The zeal for anticipating and regulating the future can be understood as part and parcel of regulatory bioethics' ascendency as a form of expertise in a risk society – in which activities in the present generate uncertainties regarding the future, fuelling an inescapable preoccupation with the future that is brought to bear on all present decisions (Giddens, 1999). In a risk society, technocratic expertise becomes vested with power owing to its perceived ability to allow for control (over nature, society, the economy, etc.) while minimizing uncertainty. Anticipatory bioethics functions accordingly: rather than allowing technological expectations to generate imaginaries abounding in uncertainty and open to competing analyses (as is the case for prophetic bioethics), it generates sociotechnical imaginaries of greater certainty predicated on bioethical guidance. This simultaneously legitimates bioethics' place in generating policy advice and legitimates the technological futures it examines (Hedgecoe, 2010; Jasanoff, 2005).

If ignoring the future could risk undermining bioethics' legitimacy, why can't authors at least moderate rather than amplify expectations? When expectations are deflated, the future begins to look very similar to the present, and the need for ethical discussions ahead of the technological curve dissipates. Bioethics tends to confer legitimacy on the technological futures it discusses even when they are highly speculative because: (i) the originality of its contribution hangs on the uniqueness of the future technology; (ii) the utility of the discourse hinges on the future existence of that particular technology; and (iii) the need to have a discussion now is predicated on the nearness of that technological future (cf. Hedgecoe, 2010). Together, these generate a strong bias towards depicting specific technological futures as departures from the present that are real, inevitable and impending. Regardless of the particulars of the discussion, which ethical issues are identified and how they are addressed, the technological future is brought into the present and reified.

Fundamentally, anticipatory ethics serves to legitimate expectations because it applies the discursive framing of regulatory bioethics prematurely. Regulatory bioethics may be useful when asking 'what do we do with the technologies that we already have?' but transposing this question to the future ('what will we do with the technologies that we will have?') evades the issue of what technologies we should have and why – the very questions relevant to deliberations regarding the direction of technoscientific research. The regulatory discourse is so suffused with a sense of technological inevitability that ignoring questions about the relationship between social values, ongoing research programmes and the desirability of the imaginaries they point towards becomes a mark of realism and pragmatism (Michael, 2000). Romantics, bio-conservatives and cranky neo-Luddites rail against the future; pragmatic bioethicists deal with it.

In the recent human enhancement discourse the rhetoric of pragmatism has reached a new level. Since we are no longer discussing discrete technologies (e.g. genetic engineering), or even a category of technologies (e.g. pharmacological performance enhancers), but human enhancement writ large, it is argued that the 'enhancement enterprise' is already under way and a foregone conclusion (Buchanan, 2011b). This shifts our gaze from substantive questions of the ends that society values, towards formal or instrumental questions of means. Once technological ends such as human biomedical enhancement are faits accomplis, then the only question that warrants attention is how we should regulate the various means that will get us there. Anticipatory bioethics has thus come to resemble areas of policy studies like technology assessment and forecasting that deal in 'instrumentally rational representations of the future', where 'the critical focus of an instrumentally rational technique such as technological foresight is primarily upon means, processes by which to reach an end which remains largely "uninterrogated"' (Michael, 2000, p. 28). Left with only a formally/instrumentally rational evaluative vocabulary, one can only focus on the application of a technology towards already accepted ends. Questions regarding ultimate ends or whether a given technology ultimately functions as an end unto itself are structurally excluded (Evans, 2002).

Simulated futures and moral agency

Regulating technological futures serves to legitimate and promote particular imaginaries by stabilizing them within an influential expert discourse – one that has a special claim to be advancing an ethical perspective. This communicates to the public that various ethical and social consequences of expected technologies are being given due regard and that technoscience is therefore not 'out of control'. Conferring legitimacy on certain imaginaries while excluding competing visions can itself contract the potential imaginative space of collective agency. However, what renders speculative bioethics truly prejudicial is the effect of looking towards the imagined future as a pressing source of moral concern, rather than at the constellation of current decisions, prioritizations and promises we face today. This reifies the possible future to the extent that it can overtake the present (Nordmann, 2007). Once the imagined future is perceived as inevitable and posing urgent questions, the range of current possibilities open to agency, deliberation and democratic input narrows; other options, prioritizations and decisional junctures can be obscured and even forgotten.

Reification allows the imagined future to overtake the present because it subverts the normal interplay between risk-driven rational agency and the indeterminacy of the future mediated by predictive models. When we consider the possible ramifications of present decisions, what is real (the present action) is set off from what is only imagined (the possible future we might produce). The more we can accurately model the effects of present actions into the future, the more 'real' these futures become, limiting our range of rationally acceptable choices. In many cases this is highly desirable because the increased 'reality' of the future forces us to take necessary actions we might otherwise resist. The example we are most familiar with today is modelling the relationship between carbon emissions and climate change, which bears significantly on present socio-political decisions precisely because these decisional possibilities can themselves be incorporated into the model.

However, in anticipatory bioethics the imagined future is not modelled as the potential outcome of various present decisions; quite the opposite: it is disconnected from the present and unresponsive to it. The model simulates only the interactions between technologies of the future, ethical principles for the future, and societal choices in the future. Within the anticipatory discourse, agency is therefore no longer located in the present but transposed into the simulation. The present recedes and the simulation becomes more real. This 'hyperreal' quality of simulations is described by Baudrillard as an increasingly ubiquitous feature of the present culture.

> Models no longer constitute an imaginary domain with reference to the real; they are, themselves, an apprehension of the real, and thus leave no room for any fictional extrapolation – they are immanent, and therefore leave no room for any kind of transcendentalism. The stage is now set for simulation, in the cybernetic sense of the word – that is to say, for all kinds of manipulation of these models (hypothetical scenarios, the creation of simulated situations, etc.), but now *nothing distinguishes this management-manipulation from the real itself: there is no more fiction.*
>
> (Baudrillard, 1991; emphasis in original)

The imagined future, now immanent and transformed into a self-contained manipulable model, overwhelms the actual present and 'leaves no room for any sort of fictional extrapolation' that would allow us to expand the space of present agency by imagining the future as otherwise.

Anticipatory bioethics relocates agency from the present to an imagined future which lies outside of time but demands our present attention. There, ethical principles and guidelines developed for that future become active, and through them the imagined future is manipulated and controlled. The imagined future is thereby transformed from a simple prediction, a promise or a sociotechnical imaginary into a simulation of the future; a hyperreal simulacrum of idealized moral agency that displaces the now-inert present in the moral imagination. The remaining function of the present is to incorporate the future to achieve a new preparatory form of agency that is not bounded by the normal epistemic limitations that constrain rationality.[21]

With the imagined future no longer a provisional stand-in for the future but its prefiguring, the logic of moral argumentation shifts as well. Arguments structured as 'if in the future we have technology X, then in principle we would have obligation P' are transformed by the inevitability factor attached to the technological antecedent. Once the future is real, it now follows that 'since we have obligation P in principle, then we ought to develop technology X to actualize P' (much in the same way that if we – in principle – have an obligation to provide sanitary living conditions, it follows that we ought to build sewers and water treatment facilities). What were hypothetical discussions of possible technologies effectively become technological imperatives that would require us to advance current research programmes toward their realization.[22]

Finally, the hyperreal future displaces the present, for it is now on a technological–decisional continuum with the present such that decisions regarding present technologies must accommodate or take into account the reality of future technologies. This can take the form of arguments that call for principles governing the entire spectrum of technologies – both current and future – falling under the rubric of human enhancement (Buchanan, 2011b), or attempts to subsume present and future technologies (such as present-day prenatal genetic selection and possible future genetic interventions) within an evaluative framework that offers a coherent perspective on both (Buchanan *et al.*, 2000). In sum, the imagined future imposes itself on the present by demanding that we interpret the present in light of its existence and act accordingly,[23] reproducing a cultural moment in which rapid technological and social changes have conditioned us in such a way that 'our present has in actuality come increasingly to make sense less as a continuation of the past than as an anticipation of the future, which it pre-empts or incorporates before it can ever arrive' (Benison, 1992, p. 158).

Reconstructing speculative bioethics

Prophetic bioethics and uncertainty

Although speculative bioethics is problematic in many respects, eliminating it altogether by focusing only on the present and ignoring the multiplicity of imagined futures associated with ongoing technoscientific research is not viable. Imagined futures will always 'intrude' into discussions of current biotechnologies, for, as STS scholarship continues to show, they play a crucial role in how scientific research is coordinated, how new technologies are developed and how we collectively envision future societies. The institutional acceptance that bioethics gained through its coordination with biomedical research and its embrace of the scientific imagination has, at least in the United States, uniquely positioned it to play a role in mediating between scientific communities and the public, and to engage researchers, policymakers and the broader public in conversations concerning the long-term aims and implications of current research programmes. Doing so without engaging with sociotechnical imaginaries would be impossible.[24]

The question is whether bioethics possesses the appropriate resources for such a task. Prophetic bioethics, concerned with 'the social and cultural meaning of the biomedical developments' (Callahan, 1999, p. 280), is certainly capable of analysing sociotechnical futures, the values implicit in their construction and their conceptual suppositions. Properly formulated, the coherence of a speculative discussion does not depend on whether the technology ever comes to fruition. Rather, its value lies in helping to clarify what kinds of futures society wishes to pursue and what futures it may already be pursuing. By bringing these considerations into public discussions, speculative bioethics could contribute to understanding decisions being made on an ongoing basis in terms of how they reflect underlying value commitments and orientations towards particular imaginaries.

Although this mode of engagement with speculative technologies is valuable, it must be accompanied by a correspondingly appropriate framing of speculative futures. Unfortunately, even when the intent is not to regulate the future and the analysis is not contingent on the realization of a specific sociotechnical future, the rhetoric of nearing radical change is often included anyway to justify attending to possibilities that could otherwise be mistaken for mere thought experiments.[25] This may be less an attempt to justify one's scholarship as relevant to others than a reflection of how those who work in applied ethics see themselves in relation to those who do more theoretical work. Nonetheless, it injects a sense of technological inevitability into the discussion, and since the object of analysis or the motivation for having the discussion is framed in terms of *the* future (or in some cases, the future we will end up with unless we significantly change course), the imagined future again threatens to overwhelm the present, with consequences for how agency is structured and how choices are prioritized.

What is missing is a way of harmonizing two seemingly conflicting requirements: to take speculative futures seriously *and* to invest them with fairly low degrees of certainty, so that they are not mistaken for actual futures or sufficiently felicitous representations thereof. What would it entail for bioethicists to maintain healthy scepticism regarding technoscientific promises and the machinery of hype, yet still justify devoting intellectual capital towards analysing them? For social scientists, this is no challenge at all, for they are interested in the shape of the discourse, warts and all. Rather than asking ethicists to assume the less engaged stance of the social scientist, I suggest fostering a closer alignment with the uncertainty internal to the practice of science itself; an appreciation of the way in which images of the future are necessary for the advancement of science, yet at the very same time remain utterly provisional, as was elegantly expressed by François Jacob:

> Our imagination displays before us the ever changing picture of the possible. It is with this picture that we incessantly confront what we fear and what we hope. It is to this possible that we adjust our wishes and our loathings. Yet, while it is part of our nature to produce a future, the system is geared in such a way that our predictions have to remain dubious. We cannot think of ourselves without a following instant, but we cannot know what this instant will be like. What we can guess today will not be realized. Change is bound to occur anyway, but the future will be different from what

we believe. This is especially true in science. The search for knowledge is an endless process and one can never tell how it is going to turn out. Unpredictability is the nature of the scientific enterprise.

(Jacob, 1982, p. 67)

Within this idealized self-conception, science expects and even desires its understanding of the world and imagining of the future to be superseded. Unfortunately, this is often obscured in the age of technoscience in which the aims of understanding the world and manipulating it to specific ends have become intertwined. Whereas the fulfilment of expectations within science lie in their unfolding and revealing new and unforeseeable possibilities – not in coming true just as anticipated – the same is not true for technological expectations. Technologies that do not deliver on their promises or work as designed are usually marked as failures. As a field, bioethics could encourage norms of discourse based on the idea that discussions pegged to particular technological futures rather than to the ongoing unfolding of new possibilities will inherit a similar tendency to turn into failures when expectations remain unfulfilled.

Towards an ethics of wishing

A constructive form of speculative bioethics would likewise understand that the sociotechnical futures that so fascinate us are also the ones that lie well beyond the veil of technical and social indeterminacy. The practical ethical questions surrounding new and emerging technologies can therefore only be addressed stepwise as research programmes are proposed and technological possibilities are developed and become realized in artefacts and techniques. However, it would also recognize that images of technological futures deserve special moral scrutiny as organizing features of the contemporary world. Prophetic approaches within bioethics have long been attuned to the 'relationships between norms and the narratives within which they are embedded' (Campbell, 1997, p. 17), and therefore already lend themselves to analysing how imagined futures shape our understanding of the present. Bioethics would ideally serve the public interest by engaging with sociotechnical imaginaries precisely to interrogate how certain imagined futures 'not only help to reconfigure actors' sense of the possible spaces of action, but also their sense of the rightness of action' (Jasanoff, 2015, p. 23). Do particular imaginaries expand the range of possibilities that appear as viable futures or narrow them into a constrained set of technological imperatives?

The idea of analysing speculative futures in such a manner has been suggested by Alfred Nordmann in his critique of the aspirations of technology assessment, which, much like anticipatory bioethics, attempts to foretell and shape the future and fails for many of the reasons discussed earlier (2010). Nordmann suggests that the key error of technology assessment is conceptualizing the future in technoscientific terms as the realization of technological projects. The remedy is to understand sociotechnical futures as inherently ahistorical. These images do not aim to represent or simulate the future; they are wishes for the future that depict the unfolding of potentialities perceived as embedded in the present. In place of technology assessment, a 'forensics of wishing' focuses not on the future but on 'cultural imaginaries and stereotypes that define research agendas and shape ideas of technical solutions to societal problems' (Nordmann, 2010, p. 13).

Bioethicists drawn to the sociotechnical scenarios that suffuse discussions of biotechnology might likewise pursue an 'ethics of wishing' – a prophetic mode of engagement that avoids reifying the future by being of and for the present. It abandons the language and pretence of predicting, representing or preparing for speculative futures, and instead asks what conceptions of the good, what shared values and what notions of human flourishing technological wishes embody. It reflects on the narratives by which these wishes are woven into larger imaginaries and asks what these stories tell us about ourselves. It looks beyond the surface appeal of a technological vision and asks what human needs it addresses and whether it offers a compelling solution compared to alternatives and in light of other, perhaps more pressing, concerns. Ethicists working on such a project would find the STS literature that traces the development and maps the dynamics of various sociotechnical imaginaries to offer vital points of departure for moral reflection and analysis. The encounter with STS would also serve to make ethicists more self-aware and sensitive to how their own discursive practices interact with and help to produce cultural imaginaries.

The future imagined and the future that achieves materiality

In this final section, I would like to briefly consider what happens to speculative bioethics discourse reconstructed in a prophetic mode when a possible future technology is realized in some form; does it become irrelevant, or does it still have something valuable to contribute to the ongoing work of understanding, monitoring and, to the extent possible, controlling the effects of newly developed biotechnologies? The discussion thus far should already give us some sense that this is a facet of a larger question: how do the prophetic and regulatory streams of bioethics jointly contribute to the field's objectives?

In its broadest sense, prophetic bioethics is an ongoing project of questioning and refining the goals that a society wishes to achieve via the development and uses of biotechnology and asking whether the actual uses of technologies align with or transmute these goals. It therefore naturally ranges over currently available technologies, those that are emerging from current research programmes and those that exist solely as either desired or angst-inducing aspects of collectively imagined futures. The challenge is to do this kind of work without conflating the real and the imagined, and to thus confuse interrogating future possibilities and grappling with actual technologies. Evaluating newly developed technologies as if they continue to exist as pure projections offers only an equally fallacious obverse of an anticipatory bioethics that tries to regulate the future long before it arrives.

What then is the value of earlier speculative discussions once a technology is in some way realized and the inevitable disjunction between an imagined sociotechnical future and an actual socially embedded technology becomes apparent? A properly configured speculative bioethics has the potential to engage with a future technology not simply as a conceivable technique and possible future artefact, but as an ongoing and evolving discourse 'of and about the technique, which fits it into a system with other techniques or know-how, with symbolic or imaginary representations, with conceptions of the world, but also with institutions, rules and norms' (Dupuy, 2007, p. 241). Technological programmes, as Dupuy notes, therefore come to fruition already carrying with them discursive elements that prefigured and helped to shape their material existence. Despite the inability to clearly foresee the social and ethical ramifications of new technologies, ethicists who are sensitive to discursive continuities that persist within technoscientific programmes may be better attuned to the ethical problems that are revealed as new technologies are implemented.

In vitro fertilization (IVF), a technology whose social consequences and ethical significance are still unfolding today, offers an example of how such a possibility may have presented itself but went largely unrealized. Before it was achieved successfully, IVF raised two fundamental bioethical issues: the regulatory question of safety for mother and future child, and prophetic/anthropological concerns about the transformation of human reproduction into a dehumanized process of manufacture – too often presented in terms of the fictional spectre of baby hatcheries from Aldous Huxley's *Brave New World* (Ramsey, 1970a). Both questions were largely thought to have been resolved as one normal baby, followed by many more, emerged from a mother's womb into ordinary families (Brody, 2009). Yet, even as IVF gained widespread acceptance as a means of circumventing fertility problems, its expanded use has also facilitated new forms of commercialized reproduction that increasingly follow some of the same trajectories as globalized industrial manufacturing, with citizens of richer countries outsourcing surrogacy to locations where it can be bought for less.

If bioethicists troubled by the potential for IVF to resemble manufacture had proceeded with a nuanced version of their critique and continued to pay close attention to the expanding applications of the technology, they might have been the first to express legitimate concerns that their worries were to some extent being realized. Whether or not this would have made any measurable impact on the way that IVF has been discussed, let alone regulated (or essentially unregulated, as is the case in the US), is impossible to say. Nonetheless, going forwards, both ethicists and STS scholars who examine emerging technologies may find that their work benefits from an alertness to the potential for the imagination – or, rather, a multiplicity of conflicting, dissenting and collectively ambivalent *imaginings* – to highlight discursive features of technologies that may become salient in the course of time.

Notes

1 Aside from numerous articles in bioethics literature, the number of books devoting significant attention to biomedical enhancement has multiplied. A sample of those published between 2000–2012 includes: Agar (2004, 2010); Buchanan (2011a, 2011b); Buchanan *et al.* (2000); Glover (2006); Gordijn and Chadwick (2008); Green (2007); Mehlman (2009); Parens (2000); Savulescu and Bostrom (2009); Savulescu,

ter Meulen and Kahane (2011); Stock (2002); Stock and Campbell (2000); Tirosh-Samuelson and Mossman (2012); and Zonneveld, Dijstelbloem and Ringoir (2008).

2 Outright scepticism regarding the expectation that genetic or other very powerful enhancements will be realized is unusual in bioethics literature, presumably because it seems unwise to bet against technoscience achieving this future given enough time. However, there are authors who expressly prefer to analyse enhancement in terms of current technologies without entangling them with future possibilities (Jones, 2006; Murray, 2007).

3 On this point it is necessary to differentiate between bioethics and the more recent 'nanoethics' (though for other purposes, such as mapping the speculative content of the human enhancement discourse, they can be aggregated). Beginning in the 1960s, bioethics, with its speculative side, developed among theologians, lawyers, clinicians and applied moral philosophers who, as I describe in the next section, saw their field as primarily devoted to real-world issues. Nanoethics emerged in the 2000s with a significant degree of participation from scholars associated with science and technology studies, and early on it was self-conscious about its speculative nature (Gordijn, 2005). The very first issue of the journal *NanoEthics* even included a critique of the entire enterprise of speculative ethics (Nordmann, 2007). Thus, while methodological questions have been explored at length by the community of scholars associated with nanoethics (Coenen and Simakova, 2013; Ferrari, Coenen and Grunwald, 2012; Grunwald, 2010; Rip and te Kulve, 2008), the same has not occurred within bioethics (see the following note).

4 Critiques first appeared from 2004–2006 as the human enhancement was becoming a central concern, primarily arguing that speculative futures were diverting attention from more serious issues and distorting how others were analysed (Elliott, 2005; Guyer and Moreno, 2004; Jones, 2006). After the emergence of nanoethics brought a critical STS perspective to speculative ethics (see previous note) some bioethicists began to incorporate it into their critiques (Jones, Whitaker and King, 2011). More recently this analytical framing has begun to be applied within bioethics in a more systematic manner (Racine *et al.*, 2014; Schick, 2014). An assessment of forward-looking bioethics can also be found in (Brody, 2009, Chapter 11).

5 Ample STS literature discusses bioethics as a legitimating and normalizing discourse on new and emerging biotechnologies, (Brosnan, 2011; Conrad and De Vries, 2011; Hedgecoe, 2010; Hedgecoe and Martin, 2008; Jasanoff, 2005; Petersen, 2011a, 2011b), but it does not differentiate between bioethical explorations of technologies expected in the relatively near term and speculative bioethics as I define it here, concerned with technological futures that are depicted as radically different from the present. Nor does it differentiate between discursive modes of bioethics, as I do below.

6 Jasanoff and Kim define sociotechnical imaginaries as 'collectively held, institutionally stabilized, and publicly performed visions of desirable futures, animated by shared understandings of forms of social life and social order attainable through, and supportive of, advances in science and technology' (Jasanoff, 2015, p. 4).

7 Different perspectives on the formation of bioethics can be found in several histories (Fox and Swazey, 2008; Jonsen, 1998; Rothman, 1991). What follows is a composite, largely following Stevens (2000) in contextualizing bioethics' emergence in a climate of widespread ambivalence and apprehensiveness regarding new technologies, and the complementary sociological/historical work by Evans (2002, 2012) that analyses how bioethics discourse developed.

8 This relationship soured within a few years. The 1975 Asilomar conference on recombinant DNA research was organized as an attempt at self-regulation without significant participation by non-scientists. However, Asilomar also differed in that it focused only on issues of risk and safety, setting aside the larger social and ethical issues that scientists and ethicists had looked at earlier (Hurlbut, 2015).

9 The reference is to Nobel laureate Jacques Monod's book *Chance and Necessity* (1972).

10 Robert Sinsheimer, a noted researcher in the area of nucleic acids, presented on the speculative future of genetics. He would later become one of the initiators of the Human Genome Project (Kevles and Hood, 1993, p. 18).

11 In my view Callahan uses the terms 'prophetic' and 'social and cultural' bioethics in ways that make them close to being interchangeable. I adopt 'prophetic' as the converse of 'regulatory' primarily due to their recurring (though inconsistent) pairing within bioethics literature (Anderlik, 2005; Callahan, 1993; Campbell, 1997; Gustafson, 1990; Guyer and Moreno, 2004; Moreno, 2005), as well as by those writing about bioethics from outside the field (Hedgecoe, 2010). The polysemy of the term 'prophetic' captures the admixture of envisioning the future, social critique and eschatology often found in speculative bioethics. However, this introduces a potential confusion, for prophetic bioethics, in the usage I adopt from Callahan, refers only to a mode of discourse – one that can be applied to the present or the future. Speculative bioethics has accordingly been pursued in both prophetic and regulatory modes, as I discuss below. Those who find the use of the term 'prophetic' unclear may wish to substitute 'social and cultural' in its place.

12 Callahan borrowed the term 'prophetic' from its use describing social activist clergy and theologians of the 1960s and 1970s. Their ethos had inflected the early stages of bioethics, in which many of the participants (Callahan included) came from backgrounds in theology (cf. Anderlik, 2005).

13 The ELSI program of the National Human Genome Research Institute (formerly the National Center for Human Genome Research) was part of the Human Genome Project nearly from its inception, apparently at the behest of James D. Watson, who initially headed the HGP (Cook-Deegan, 1994). Three to 5 per cent of the HGP's annual budget was put towards ELSI, which amounted to about $162 million between 1991 and 2003 (Bennett and Sarewitz, 2006).

14 Even as this source of funding was welcomed, it was soon criticized by some ethicists for allowing hype to skew attention towards genethics (Annas, 1992) and producing redundant, low-quality research (Turner, 2003).

15 This reflects a wider cultural moment. 1997 saw the publication of a scientist's science-fictional treatise on the anticipated effects that genetic enhancement would have on future societies (Silver, 1997), and the release of the science fiction movie *GATTACA* (Niccol, 1997), depicting a dystopian version of one such future. It was also the year in which the birth of Dolly the sheep was announced, bringing speculative futures closer still.

16 As is indicated in the book's introduction, some portion of the project funding came from the ELSI programme (Buchanan *et al.*, 2000, p. xiv).

17 This was a prophetic discourse in that it brought competing substantive views regarding enhancement to the discussion and was interested in how technology affects the human condition. However, what distinguished this particular development was that it reintroduced a visionary style of utopian/dystopian writing previously associated with theologians like Paul Ramsey (1970a) on the conservative side and scientists like Julian Huxley on the transhumanist side (Huxley, 1957).

Prophetic discourse in the reflective, anthropological mode that I describe had never disappeared from bioethics, it was simply far less prominent and influential. By the 1990s Callahan was attempting to move the Hastings Center and the field back in this direction (1990, 1993, 1996, 1999; see also Stevens, 2000, Chapter 2). However, his writing evinces a preference for a prophetic bioethics concerned with the ends of biomedicine generally rather than speculative futures (2000).

18 A quantitative analysis of key terms in databases that index bioethics literature shows a striking shift in terminology from 'genetic engineering' to a more general rubric of 'enhancement' (Schick, 2014, Chapter 2).

19 Reflective equilibrium is a term coined by John Rawls for a process of moral reasoning regarding basic social structures. One first identifies certain fixed points of moral conviction and then proceeds to extend these in a manner that balances them with various other competing principles. Within bioethics, reflective equilibrium is invoked far more loosely as a universal methodology of sorts for moral deliberation (Arras, 2007). The idea of incorporating speculative futures into the process of reflective equilibrium is explicitly endorsed in (Buchanan *et al.*, 2000).

20 Note that it was 13 years after the completion of ELSI sponsored work on a draft bill for genetic privacy (Annas, Glantz and Roche, 1995) that similar federal legislation became law. Moreover, a considered analysis of the cloning controversy shows that anticipatory bioethics discourse functions quite differently before public concern has escalated (Jelsøe *et al.*, 2006).

21 Concern for the actual future – the one that will be shaped by present decisions in ways that we cannot come close to fully anticipating – entails an entirely different mode of discourse that cannot help but acknowledge the epistemic and agentic limitations of the present. It is concerned with things like urgent social and environmental problems, theories of intergenerational justice and decision-making from behind the veil of indeterminacy.

22 It is almost always apparent that authors who provide guiding principles for governing possible future technologies do so because they have already decided that having technologies of this sort would, on balance, be positive – not because they are simply open to neutral possibilities that may or may not come to be. If bioethicists truly believed, as in Agar's words, that 'we need principles for situations that might never eventuate, but whose possibility cannot be ruled out given our current state of knowledge', then one would expect to see a robust discourse on how to ration food during a global famine, how to best relocate refugees from cities receding into the oceans, how to adjudicate regional wars fought over sources of fresh water, and all sorts of worst-case global 'lifeboat ethics' scenarios that certainly cannot be ruled out.

23 Milburn notes the same form of argument in early discussions of nanotechnology (2002, p. 292).

24 Retaining speculative bioethics does not rule out greatly reducing the amount of time and effort that bioethics collectively expends on discussing future technologies.

25 To an extent, Hedgecoe's analysis of forward-looking *regulatory* mode discussions of pharmacogenomics as largely falling into the bottom (i.e. high level) of the 'certainty trough' (2010, p. 179) also holds true of the *prophetic* mode speculative enhancement discourse discussed here. I would, however, qualify this in two ways. First, there is significant variability between authors, and one might find an introductory article to an anthology using hyperbolic language (e.g. Baillie and Casey, 2004) followed by an article devoid of it. Second, as I briefly mentioned earlier, there is often hedging language that attempts to simultaneously legitimate the discussion without appearing overly committed.

All this would have to be analysed more carefully to draw conclusions about continuities between 'slightly speculative' and highly speculative bioethics, while taking into account the distinction between regulatory and prophetic modes. Once we move past regulatory bioethics, other aspects of his analysis tend not carry over, such as ethical agenda-setting by scientists and a lack of substantive arguments. Unfortunately, Hedgecoe does not make such a distinction, which would have allowed his concluding call for a form of bioethics that 'provides a rigorous, critical engagement with scientific ideas, helping policy-makers, doctors and researchers think about the kinds of futures we can live in' (2010, p. 180) to more concretely point to the already established prophetic mode of bioethics suited for this task.

References

Agar, N., 2004. *Liberal eugenics: in defence of human enhancement.* Malden, MA: Blackwell.

Agar, N., 2007. Whereto transhumanism? The literature reaches a critical mass. *Hastings Center Report*, 37(3), pp. 12–17. http://doi.org/10.1353/hcr.2007.0034.

Agar, N., 2010. *Humanity's end: why we should reject radical enhancement.* Cambridge, MA: MIT Press.

Anderlik, M.R., 2005. Respecting difference and moving beyond regulation: tasks for U.S. bioethics commissions in the twenty-first century. *Kennedy Institute of Ethics Journal*, 15(3), pp. 289–303.

Annas, G.J., 1992. The human genome project as social policy: implications for clinical medicine. *Bulletin of the New York Academy of Medicine*, 68(1), pp. 126–134.

Annas, G.J., Glantz, L.H. and Roche, P., 1995. *Genetic privacy act and commentary.* [Online] Boston: Human Genome Project Information Archive. Available at: www.ornl.gov/sci/techresources/Human_Genome/resource/privacy/privacy1.html [accessed 17 April 2013].

Arras, J.D., 2007. The way we reason now: reflective equilibrium in bioethics. In: B. Steinbock, ed. 2007. *The Oxford handbook of bioethics.* Cambridge and New York, NY: Oxford University Press. pp. 46–71.

Baillie, H.W. and Casey, T.K. eds., 2004. *Is human nature obsolete?: Genetics, bioengineering, and the future of the human condition.* Cambridge, MA: MIT Press.

Baudrillard, J., 1991. Simulacra and science fiction. *Science Fiction Studies*, 18(3), pp. 309–313.

Beauchamp, T.L. and Childress, J.F., 2008. *Principles of biomedical ethics.* 6th ed. New York, NY: Oxford University Press.

Benison, J., 1992. Science fiction and postmodernity. In: F. Barker, P. Hulme and M. Iversen, eds. 1992. *Postmodernism and the re-reading of modernity.* Manchester: Manchester University Press. pp. 138–158.

Bennett, I. and Sarewitz, D., 2006. Too little, too late? Research policies on the societal implications of nanotechnology in the United States. *Science as Culture*, 15(4), pp. 309–325. http://doi.org/10.1080/09505430601022635.

Borup, M., Brown, N., Konrad, K. and Van Lente, H., 2006. The sociology of expectations in science and technology. *Technology Analysis & Strategic Management*, 18(3–4), pp. 285–298. http://doi.org/10.1080/09537320600777002.

Brody, H., 2009. *The future of bioethics.* New York, NY: Oxford University Press.

Brosnan, C., 2011. The sociology of neuroethics: expectational discourses and the rise of a new discipline. *Sociology Compass*, 5(4), pp. 287–297.

Brown, N., 2003. Hope against hype: accountability in biopasts, presents and futures. *Science Studies*, 16(2), pp. 3–21.

Brown, N. and Michael, M., 2003. A sociology of expectations: retrospecting prospects and prospecting retrospects. *Technology Analysis & Strategic Management*, 15(1), pp. 3–18.

Brown, N., Rappert, B. and Webster, A. eds., 2000. *Contested futures: a sociology of prospective science and technology.* Aldershot: Ashgate.

Buchanan, A., 2011a. *Better than human: the promise and perils of enhancing ourselves.* New York, NY: Oxford University Press.

Buchanan, A., 2011b. *Beyond humanity?: The ethics of biomedical enhancement.* New York, NY: Oxford University Press.

Buchanan, A., Brock, D.W., Daniels, N. and Wikler, D., 2000. *From chance to choice: genetics and justice*. Cambridge; New York, NY: Cambridge University Press.

Callahan, D., 1973. Bioethics as a discipline. *The Hastings Center Studies (Report)*, 1(1), pp. 66–73. http://doi.org/10.2307/3527474.

Callahan, D., 1990. Religion and the secularization of bioethics. *The Hastings Center Report*, 20(4), pp. 2–4. http://doi.org/10.2307/3562776.

Callahan, D., 1993. Why America accepted bioethics. *The Hastings Center Report*, 23(6), pp. S8–S9.

Callahan, D., 1996. Bioethics, our crowd, and ideology. *The Hastings Center Report*, 26(6), pp. 3–4. http://doi.org/10.2307/3528741.

Callahan, D., 1999. The social sciences and the task of bioethics. *Daedalus*, 128(4), pp. 275–294.

Callahan, D., 2000. Judging the future: whose fault will it be? *Journal of Medicine and Philosophy*, 25(6), pp. 677–687. http://doi.org/10.1076/jmep. 25.6.677.6127.

Campbell, C.S., 1997. Prophecy and policy. *Hastings Center Report*, 27(5), pp. 15–17. http://doi.org/10.2307/3527795.

Chapman, A.R. and Frankel, M.S. eds., 2003. *Designing our descendants: the promises and perils of genetic modifications*. Baltimore, MD: The Johns Hopkins University Press.

Clayton, J., 2013. The ridicule of time: science fiction, bioethics, and the posthuman. *American Literary History*, 25(2), pp. 317–340.

Coenen, C. and Simakova, E., 2013. STS policy interactions, technology assessment and the governance of technovisionary sciences. *Science, Technology & Innovation Studies*, 9(2), pp. 3–20.

Collingridge, D., 1980. *The social control of technology*. New York, NY: St. Martin's Press.

Condliffe, P. and Callahan, D., 1973. Preface. In: B. Hilton, D. Callahan, M. Harris, P. Condliffe and B. Berkley, eds. 1973. *Ethical issues in human genetics*. New York, NY: Plenum Press. pp. ix–xi.

Conrad, E.C. and De Vries, R., 2011. Field of dreams: a social history of neuroethics. *Advances in Medical Sociology*, 13, pp. 299–324. http://doi.org/10.1108/S1057-6290(2011)0000013017.

Cook-Deegan, R., 1994. *The gene wars: Science, politics, and the human genome*. New York, NY: Norton.

Dupuy, J.-P., 2007. Some pitfalls in the philosophical foundations of nanoethics. *Journal of Medicine and Philosophy*, 32(3), pp. 237–261. http://doi.org/10.1080/036053107013 96992.

Elliott, C., 2005. Adventure! Comedy! Tragedy! Robots! How bioethicists learned to stop worrying and embrace their inner cyborgs. *Journal of Bioethical Inquiry*, 2(1), pp. 18–23.

Etzioni, A., 1973. *Genetic fix*. New York, NY: Macmillan.

Evans, J.H., 2002. *Playing God?: Human genetic engineering and the rationalization of public bioethical debate*. Chicago, IL, and London: University of Chicago Press.

Evans, J.H., 2012. *The history and future of bioethics: a sociological view*. New York, NY: Oxford University Press.

Ferrari, A., Coenen, C. and Grunwald, A., 2012. Visions and ethics in current discourse on human enhancement. *NanoEthics*, 6(3), pp. 215–229. http://doi.org/10.1007/s11569-012-0155-1.

Fletcher, J.F., 1974. *The ethics of genetic control: ending reproductive roulette*. Garden City, NY: Anchor.

Fox, R.C. and Swazey, J.P., 2008. *Observing bioethics*. New York, NY: Oxford University Press.

Fukuyama, F., 2002. *Our posthuman future: consequences of the biotechnology revolution*. New York, NY: Farrar, Straus and Giroux.

Gattaca, 1997. [Film] Directed by Andrew Niccol. USA: Columbia Pictures, Jersey Films.

Giddens, A., 1999. *Runaway world – Reith lectures*. [Online] available at: www.bbc.co.uk/radio4/reith1999/lecture2.shtml [accessed 30 June 2016].

Glover, J., 2006. *Choosing children: Genes, disability, and design*. Oxford and New York, NY: Oxford University Press.

Goodfield, J., 1977. *Playing God: genetic engineering and the manipulation of life*. New York, NY: Random House.

Gordijn, B., 2005. Nanoethics: from utopian dreams and apocalyptic nightmares towards a more balanced view. *Science and Engineering Ethics*, 11(4), pp. 521–533.

Gordijn, B. and Chadwick, R. eds., 2008. *Medical enhancement and posthumanity*. Dordrecht: Springer.

Green, R.M., 2007. *Babies by design: the ethics of genetic choice*. New Haven, CT: Yale University Press.

Grunwald, A., 2010. From speculative nanoethics to explorative philosophy of nanotechnology. *NanoEthics*, 4(2), pp. 91–101. http://doi.org/10.1007/s11569-010-0088-5.

Gustafson, J.M., 1990. Moral discourse about medicine: a variety of forms. *Journal of Medicine and Philosophy*, 15(2), pp. 125–142.

Guyer, R.L. and Moreno, J.D., 2004. Slouching toward policy: lazy bioethics and the perils of science fiction. *The American Journal of Bioethics*, 4(4), pp. W14–17. http://doi.org/10.1080/15265160490908022.

Hedgecoe, A., 2010. Bioethics and the reinforcement of socio-technical expectations. *Social Studies of Science*, 40(2), pp. 163–186.

Hedgecoe, A. and Martin, P.A., 2008. Genomics, STS, and the making of sociotechnical futures. In: E.J. Hackett, ed. 2008. *The handbook of science and technology studies*. 3rd ed. Cambridge, MA: MIT Press. pp. 817–840.

Hilton, B., Callahan, D., Harris, M., Condliffe, P. and Berkley, B. eds., 1973. *Ethical issues in human genetics*. New York, NY: Plenum.

Hood, L., 2000. The human genome project – launch pad for human genetic engineering. In: G. Stock and J.H. Campbell, eds. 2000. *Engineering the human germline*. New York, NY: Oxford University Press. pp. 17–24.

Hurlbut, J.B., 2015. Limits of responsibility: genome editing, asilomar, and the politics of deliberation. *Hastings Center Report*, 45(5), pp. 11–14. http://doi.org/10.1002/hast.484.

Huxley, J., 1957. *New bottles for new wine*. London: Chatto & Windus.

Jacob, F., 1982. *The possible and the actual*. New York, NY: Pantheon.

Jasanoff, S., 2005. *Designs on nature: science and democracy in Europe and the United States*. Princeton, NJ: Princeton University Press.

Jasanoff, S., 2015. Future imperfect: science, technology, and the imaginations of modernity. In: S. Jasanoff and S.-H. Kim, eds. 2015. *Dreamscapes of modernity: sociotechnical imaginaries and the fabrication of power*. Chicago, IL, and London: University of Chicago Press. pp. 1–33.

Jasanoff, S. and Kim, S.-H. eds., 2015. *Dreamscapes of modernity: sociotechnical imaginaries and the fabrication of power*. Chicago, IL, and London: University of Chicago Press.

Jelsøe, E., Mortensen, A.T., Kamara, M.W., Rusanen, M., Costa, S., Nielsen, T.H. and Lindsey, N., 2006. Moving the goalposts in bioethics. In: G. Gaskell and M.W. Bauer, eds. 2006. *Genomics and society: legal, ethical, and social dimensions*. London and Sterling, VA: Earthscan. pp. 44–59.

Jones, D.G., 2006. Enhancement: are ethicists excessively influenced by baseless speculations? *Medical Humanities*, 32(2), pp. 77–81. http://doi.org/10.1136/jmh.2005.000234.

Jones, D.G., Whitaker, M. and King, M., 2011. Speculative ethics: valid enterprise or tragic cul-de-sac? In: A. Rudnick, ed. 2011. *Bioethics in the 21st century*. InTech. Available at: www.intechopen.com/books/bioethics-in-the-21st-century/speculative-ethics-valid-enterprise-or-tragic-cul-de-sac- [accessed 24 June 2013].

Jonsen, A.R., 1998. *The birth of bioethics*. New York, NY: Oxford University Press.

Jotterand, F., 2010. At the roots of transhumanism: from the enlightenment to a post-human future. *Journal of Medicine and Philosophy*, 35(6), pp. 617–621. http://doi.org/10.1093/jmp/jhq050.

Juengst, E.T., 1991a. Germ-line gene therapy: back to basics. *Journal of Medicine and Philosophy*, 16(6), pp. 587–592.

Juengst, E.T., 1991b. The human genome project and bioethics. *Kennedy Institute of Ethics Journal*, 1(1), pp. 71–74.

Kevles, D.J. and Hood, L.E., 1993. *The code of codes: scientific and social issues in the human genome project*. Cambridge, MA: Harvard University Press.

Khushf, G., 2007. Open questions in the ethics of convergence. *Journal of Medicine and Philosophy*, 32(3), pp. 299–310. http://doi.org/10.1080/03605310701397057.

Kramer, P.D., 1993. *Listening to Prozac: a psychiatrist explores antidepressant drugs and the remaking of the self.* 1st ed. New York, NY: Viking Adult.

Lin, P. and Allhoff, F., 2006. Nanoethics and human enhancement: a critical evaluation of recent arguments. *Nanotechnology Perceptions*, 2(1), p. 47.

McLaren, A., 2012. *Reproduction by design: sex, robots, trees, and test-tube babies in interwar Britain*. Chicago, IL, and London: University of Chicago Press.

Mehlman, M.J., 2009. *The price of perfection: individualism and society in the era of biomedical enhancement*. Baltimore, MD: Johns Hopkins University Press.

Michael, M., 2000. Futures of the present: from performativity to prehension. In: N. Brown, B. Rappert and A. Webster, eds. 2000. *Contested futures: A sociology of prospective science and technology*. Aldershot: Ashgate. pp. 21–39.

Milburn, C., 2002. Nanotechnology in the age of posthuman engineering: science fiction as science. *Configurations*, 10(2), pp. 261–295. http://doi.org/10.1353/con.2003.0017.

Mnyusiwalla, A., Daar, A.S. and Singer, P.A., 2003. 'Mind the gap': science and ethics in nanotechnology. *Nanotechnology*, 14(3), R9–R13. http://doi.org/10.1088/0957-4484/14/3/201.

Monod, J., 1972. *Chance and necessity: an essay on the natural philosophy of modern biology* (A. Wainhouse, trans.). New York, NY: Vintage.

Moreno, J.D., 1995. *Deciding together: bioethics and moral consensus*. Oxford: Oxford University Press.

Moreno, J.D., 2005. The end of the great bioethics compromise. *Hastings Center Report*, 35(1), pp. 14–15. http://doi.org/10.1353/hcr.2005.0011.

Murray, T.H., 2007. Enhancement. In: B. Steinbock, ed. 2007. *The Oxford handbook of bioethics*. Oxford and New York, NY: Oxford University Press. pp. 491–515.

Nelkin, D. and Lindee, M.S., 2004. *The DNA mystique: the gene as a cultural icon*. 2nd ed. Ann Arbor, MI: University of Michigan Press.

Nordmann, A., 2007. If and then: a critique of speculative nanoethics. *NanoEthics*, 1(1), pp. 31–46. http://doi.org/10.1007/s11569-007-0007-6.

Nordmann, A., 2010. A forensics of wishing: technology assessment in the age of techno-science. *Poiesis & Praxis*, 7(1–2), pp. 5–15. http://doi.org/10.1007/s10202-010-0081-7.

Parens, E., 1998. Is better always good?: The enhancement project. *Hastings Center Report*, 28(1), pp. s1–s17. http://doi.org/10.2307/3527981.

Parens, E., 2000. *Enhancing human traits: ethical and social implications.* Washington, DC: Georgetown University Press.

Petersen, A., 2011a. Can and should sociology save bioethics? *Medical Sociology Online*, 6(1), pp. 2–14.

Petersen, A., 2011b. *The politics of bioethics.* New York, NY, and London: Routledge.

President's Commission for the Study of Ethical Problems in Medicine and Biomedical and Behavioral Research, 1982. *Splicing life: a report on the social and ethical issues of genetic engineering with human beings.* Washington, DC.

Racine, E., Rubio, T.M., Chandler, J., Forlini, C. and Lucke, J., 2014. The value and pitfalls of speculation about science and technology in bioethics: the case of cognitive enhancement. *Medicine, Health Care and Philosophy*, 17(3), pp. 325–337. http://doi.org/10.1007/s11019-013-9539-4.

Ramsey, P., 1970a. *Fabricated man: the ethics of genetic control.* New Haven, CT: Yale University Press.

Ramsey, P., 1970b. *The patient as person: explorations in medical ethics.* New Haven, CT: Yale University Press.

Rip, A. and te Kulve, H., 2008. Constructive technology assessment and socio-technical scenarios. In: E. Fisher, C. Selin and J.M. Wetmore, eds. 2008. *The yearbook of nanotechnology in society, volume I: Presenting futures.* Dordrecht: Springer. pp. 49–70.

Roco, M.C. and Bainbridge, W.S., 2003. *Converging technologies for improving human performance: nanotechnology, biotechnology, information technology and cognitive science.* Dordrecht and Boston, MA: Kluwer Academic.

Rosenfeld, A., 1969. *The second genesis; the coming control of life.* Englewood Cliffs, NJ: Prentice-Hall.

Rothman, D., 1991. *Strangers at the bedside: a history of how law and bioethics transformed medical decision-making.* New York, NY: Basic.

Savulescu, J. and Bostrom, N. eds., 2009. *Human enhancement.* Oxford and New York, NY: Oxford University Press.

Savulescu, J., ter Meulen, R. and Kahane, G. eds., 2011. *Enhancing human capacities.* Chichester: Wiley-Blackwell.

Schick, A., 2014. *Lessons for the future? Prophecy and policy in speculative bioethics.* PhD. Michigan State University, East Lansing, MI. Available at: http://gradworks.umi.com/36/14/3614927.html [accessed 11 June 2014].

Schick, A., 2016. Whereto speculative bioethics? Technological visions and future simulations in a science fictional culture. *Medical Humanities*, 42(4), pp. 225–31. doi: 10.1136/medhum-2016-010951.

Silver, L.M., 1997. *Remaking Eden: cloning and beyond in a brave new world.* New York, NY: HarperCollins.

Squier, S.M., 1994. *Babies in bottles.* New Brunswick, NJ: Rutgers University Press.

Stevens, M.L.T., 2000. *Bioethics in America: origins and cultural politics.* Baltimore, MD: Johns Hopkins University Press.

Stock, G., 2002. *Redesigning humans: our inevitable genetic future.* New York, NY: Houghton Mifflin.

Stock, G. and Campbell, J.H. eds. 2000. *Engineering the human germline*. New York, NY: Oxford University Press.

Tirosh-Samuelson, H., 2012. Science and the betterment of humanity: three British prophets of transhumanism. In: H. Tirosh-Samuelson and K.L. Mossman, eds. 2012. *Building better humans?: Refocusing the debate on transhumanism*. Frankfurt am Main: Peter Lang. pp. 55–82.

Tirosh-Samuelson, H. and Mossman, K.L. eds., 2012. *Building better humans?: Refocusing the debate on transhumanism*. Frankfurt am Main: Peter Lang.

Toffler, A., 1970. *Future shock*. New York, NY: Random House.

Turner, L., 2003. The tyranny of 'genethics'. *Nature Biotechnology*, 21(11), pp. 1282–1282. http://doi.org/10.1038/nbt1103-1282.

Zonneveld, L., Dijstelbloem, H. and Ringoir, D. eds., 2008. *Reshaping the human condition: exploring human enhancement*. The Hague: Rathenau Institute.

3 The new biology of the social

Shaping humans' future, science, and public health

Jan Baedke[1]

This chapter deals with the recent re-emergence of the concept of the social in biology. This new so-called 'social biology' is built on a postgenomic view of developing organisms in which the material and especially social environment of organisms is thought to shape not only the expression profile of genes but also humans' ontogenetic and even transgenerational destinies. This field heavily draws on work in epigenetics on environmentally sensitive regulatory processes. Epigenetic studies in humans focus on phenomena like cancer, obesity, ageing, schizophrenia, autism and trauma, which may be induced by incisive events in early pre- or postnatal development and transmitted non-genetically across generations. This work on 'social and environmental programming' currently stirs imaginaries of human enhancement and possible future societies.

Against this background, two issues will be discussed: First (in sections 1 and 2), it will be investigated how the explanatory framework of postgenomics in general and epigenetics in particular leads to new so-called 'sociotechnical imaginaries' of humans' future. It will be shown that a number of postgenomic views of this future are grounded in a new link between biological concepts used in molecular and developmental biology (like phenotypic plasticity and organism) and social, anthropological and normative concepts (like milieu, body, identity, innateness, self-determination and responsibility). Second (in sections 3 and 4), the impact of this new conceptual synthesis between the social and the biological on science and public health policies will be discussed.

From epigenetics to the social

Currently, epigenetics is one of the 'hottest' topics in biology. The number of papers containing 'epigenetic(s)' in their title has increased tenfold since 1999. Moreover, the highly ambitious Human Epigenome Project has recently been launched. Back in the 1940s, embryologist Conrad Hall Waddington founded the field of epigenetics as the 'science concerned with the causal analysis of development' (Waddington, 1952, p. vi). At that time, epigenetics was basically developmental genetics. Throughout the decades, the causal agents of interest in epigenetics have changed, from genes to everything but – or, as some might say, everything above – genes.

Today, epigenetics primarily investigates those regulatory non-DNA factors, like DNA methylation, histone acetylation marks and small RNAs, which are taken to be responsible for whether genetic information is realized. These tiny regulating agents seem to be gradually taking over genes' explanatory supremacy. They are not only addressed in order to explain developmental phenomena, like phenotypic plasticity, but, more specifically, cancer, obesity and ageing. Moreover, epigenetic factors are highlighted in investigations of heredity phenomena, like disease aetiology, sex-linked inheritance patterns and, last but not least, the role of development in evolution.

The latter refers to the idea, defended famously by Eva Jablonka and Marion Lamb in various books and papers (see e.g. Jablonka and Lamb, 1995, 2005, 2010), that the non-random responsiveness of regulatory epigenetic factors to environmental cues as well as their heritability could drive and bias evolutionary change. In detail, this means that epigenetic variation is produced and inherited with a degree of autonomy from the DNA level. In other words, the emergence and transmission of genetic and epigenetic variation is decoupled to a certain degree. Epigenetic transmission can occur in linages of cells – *intraorganismic non-genetic inheritance* – and between organisms – *interorganismic non-genetic inheritance*. In animals, examples of the latter type of transgenerational transmission are the inheritance of a more stress-resistant phenotype in rats through extensive maternal care (Kappeler and Meaney, 2010), the enhancement of synaptic plasticity and memory formation in offspring of mice exposed to an 'enriched' environment (including novel objects, a high number of social interactions, etc.; see Arai *et al.*, 2009), male offspring of rats that show abnormalities in their testes and immune system if their pregnant mother had been exposed to endocrine disruptors (i.e. environmental chemicals, widely distributed in humans' environment; see Anway *et al.*, 2005), and an increase in body size and reduced insulin sensitivity in the offspring of female mice which had been given a high-fat diet (Dunn and Bale, 2011).

Jablonka and Lamb, among others, argue that, rather than the genes, it is the complex, environmentally responsive system of the developing organism which takes the lead in evolution. In a quasi-Lamarckian manner it non-randomly responds to environmental cues and thus introduces and transmits new phenotypes in populations which are subsequently stabilized by genes. Thus, epigenetics challenges gene-centrism and invokes a broader notion of heredity, particularly in contrast to the gene-centred view of evolution.

Recently, this new evolutionary view has been accompanied by a common call for expanding evolutionary theory towards a so-called 'extended synthesis' (e.g. Jablonka, 2006; Pigliucci and Müller, 2010; Laland *et al.*, 2015). It first offers a more complex picture of the developing organism, with various many-to-many-dependencies, rather than one-to-one-dependencies between genes and phenotypic traits, as suggested by the unidirectional model known as the 'central dogma' of molecular biology (Crick, 1958). Second, the complexity of this so-called 'genotype-phenotype-map' is increased by considering inner- and extra-organismic environmental factors as causally and explanatorily relevant.

The extended synthesis is thought to include, besides epigenetics, a number of related progressive biological fields, such as evolutionary developmental biology (evo-devo) and niche construction theory. Like epigenetics, evo-devo emphasizes the neglected role of developmental biology and embryology in the theory of evolution (e.g. articles in Sansom and Brandon, 2007; Minelli, 2009). Evo-devoists seek to explain the evolvability of developmental systems and how changes in networks of genes lead to evolutionary novelties (such as wings, eyes, etc.).[2] Niche construction theory focuses on organismic activity, especially the self-perpetuating and reciprocal effects of organisms that construct their own niche (and/or that of other species) during development (e.g. Odling-Smee, Laland and Feldman, 2003; Odling-Smee, 2010). A common example is the behaviour of beavers, which shape their ecosystem and thus alter the selection pressure acting on them and their offspring. Such cases include processes like parental effects, social learning and even symbolic communication – processes which often include epigenetic inheritance.

The new theoretical framework delineated by these fields is often described as *postgenomics* (e.g. Stotz, 2008). It states that developmental and inheritance processes are open systems in which we find multi-factorial dependencies between environmental factors, developmental mechanisms and the genome. The 'how' and 'when' of genetic information is determined not intrinsically but rather by the genes' context or environment.

Attributing an independent and important causal role to the intra- and extra-organismic context means that phenotypic traits can be (to a certain degree) *robust* during development, although the genotype might have changed. In this case, known as 'canalization', for example, genes are 'overruled' by contextual factors. In addition, many phenotypic traits are *plastic*. The concept of phenotypic plasticity holds that a trait of an organism can react to an environmental input in various ways. This means that the genome codes for a wide range of potential phenotypes. Examples of plastic traits are the morphological and behavioural changes in locust species (short-winged, uniformly coloured and solitary vs. long-winged, brightly coloured and gregarious), which are induced by changes in population density (Janzen and Phillips, 2006).

While most of the studies in epigenetics are conducted on model organisms, like mice and the worm *C. elegans*, recently the number of human studies has rapidly increased. These studies deal with the emergence and transmission of diseases like cancer, obesity, alcoholism and ageing, as well as schizophrenia, autism and trauma, which might be induced during critical periods in early life. Moreover, they seek to develop medical applications, like stem cell therapy and cloning. Often they conduct epidemiological analyses on historical records in order to investigate the intra- and transgenerational effects of particular incisive events in the biography of members of a certain social group. Such events in early pre- or postnatal life are usually exposure to stress, a particular diet or toxins during a critical period in which, for example, methylation marks in the genome show a high level of plasticity. This epigenetic variation can then be transmitted to the children (although less faithfully than genetic variation). For example, changes in small RNAs can be inherited through the placenta of the mother; or parental gene expression patterns can be re-established in every generation anew through a particular behaviour of the parents that induces these patterns. Human studies in epigenetics focus on events as diverse as violence against women during pregnancy and the transgenerational effects of the Dutch hunger winter of 1944/45 (Painter *et al.*, 2008; Radtke *et al.*, 2011).

Best known is the so-called 'Överkalix study', named after a small, isolated Swedish city (e.g. Kaati, Bygren and Edvinsson, 2002; Pembrey *et al.*, 2006). The researchers found that, from 1890 to 1920, poor food supply (due to bad crops) for boys between the ages of nine and twelve and thus during the maturation of sperm cells was associated (with high statistical significance) with a lower mortality risk and a lower risk of cardiovascular diseases in their male grandchildren. If the grandfathers had a good food supply during the same developmental period, their male grandchildren more often suffered from diabetes, had heard attacks and strokes and, on average, died earlier. Similar associations were found between grandmothers and female grandchildren. This is likely because of a sex-related inheritance of gene expression (e.g. methylation) patterns, so-called 'parental imprinting', on loci in the genome coding for the above diseases.

Often these cases are called '*social*' or '*environmental programming*' (see Meloni, 2014b). This concept refers to the basic idea that the availability of and interaction with material and social resources are directly causally related to physiological and mental properties in humans. For example, Borghol and colleagues (2012) showed that the DNA methylation profile in a cohort of adult males born during one week in 1958 in Great Britain is associated more with the – as they call it – 'socio-economic position' of these men in early life than with the adult socio-economic position. In this study, the position of the children is defined as privileged or disadvantaged based on the following criteria: the occupation of the father, lacking or sharing access to household amenities, and household overcrowding.

The associations traced in such 'sociobiological' studies often indicated a 'transgenerational vicious circle' (see Plagemann, 2012; Schuol, 2014). This means that the induced phenotype can maintain longer, even though variation in parental epigenetic marks, like DNA methylation patterns, is, if inherited (most widely), erased during the children's early development and thus transmitted mostly for only one or two generations. This is the case if an environmentally induced epigenetic variation triggers the maintenance of the – especially social – developmental niche in every generation anew. For example, the disposition of suffering from metabolic syndrome (i.e. the co-occurrence of e.g. elevated blood pressure, obesity and insulin resistance) can be environmentally induced through stress and hypercaloric food and then epigenetically inherited. This transmission triggers a 'feedback loop'. If the disposition is inherited by a girl who gets pregnant later in life, her organism, which suffers from metabolic syndrome, constitutes the imprinting context for her children. In more general terms, the social 'niche' literally gets under the skin and is stabilized by a transgenerational process of cyclic induction.

In what follows, I will not discuss the validity of these studies in epigenetics; however, I would like to remark that, today, various results in epigenetics on transgenerational epigenetic inheritance and carcinogenesis research are well-founded and supported by strong evidence. These results recently gained attention throughout the academic world through widely discussed publications in journals such as *Cell*, *Nature* and *Science*. As a consequence, epigenetics has started to impact mainstream biology as well as arouse interest in biomedicine and pharmacology (see Rodriguez-Paredes and Esteller, 2011; Arrowsmith *et al.*, 2012). Despite this success, a number of research results, especially those of human studies, are anything but fully certain. Nevertheless, these human studies in particular have stirred a multifarious and rapidly expanding discourse inside and outside of the sciences on how humans' developmental and inheritance processes reciprocally interact with their social environment. In addition, they have started exerting influence on public health applications (see section 3). Thus, today we are already being called upon to analyse the influence of epigenetic research results on our society – especially, as I will show, on individual and collective behaviours and responsibilities, as well as their justification through imaginaries of a future society. In other words, we may ask what conception of human beings and their duties in shaping our – possibly better – future is currently developed based on postgenomic approaches like epigenetics.

Sociotechnical imaginaries and the new social biology

In fact, epigenetics and studies of humans' future have a common origin: Conrad Hall Waddington. He not only coined the term 'epigenetics' but, in his later work, was among the first who tried to establish 'future studies' as an academic discipline (Waddington, 1975). He thought of this field as being concerned, first and foremost, with the 'methodologies of dealing with complex systems' (1975, p. 3), in which social actions, economical scenarios, scientific and technological innovations, etc., as well as their non-linear interactions, create complexes of problems. This includes issues of worldwide scope like nutrition and public health, overpopulation, energy supply, urbanization and so forth. Not least owing to their complexity, all of these problems go along with uncertainties, public hopes and fears, and collective views of human future.

More recently, such imagined futures and their roles in stimulating discourse, shaping policymaking and social interventions have been highlighted by Sheila Jasanoff and colleagues (Jasanoff and Kim, 2009; Jasanoff, 2011; Pickersgill, 2011). They describe them as '*sociotechnical imaginaries*'. On the one hand, these imaginaries are understood as collectively envisioned forms of social life that may be used to justify investments in science and technology and drive, bias, and structure political and/or individual decision-making. On the other hand, any sociotechnical imaginary may itself arise from scientific progress and technological innovation.

As a consequence, with respect to the issue at stake, this concept becomes crucial in investigating how epigenetics influences collective views of humans' future and thus shapes public discourse on how and why to realize this future. Below, it will be shown that the current biological discussion of the role of epigenetic mechanisms in development and evolution has led to a number of transhumanist imaginaries of the future and human enhancement in academia and the public. I call them '*epigenetic imaginaries*'. Interestingly, the basic concepts underlying these new social-utopian imaginaries are directly developed out of biologists' postgenomic conceptual framework.

Accompanied by growing attention in philosophical anthropology, bioethics, science and technology studies and social theory, the amount of popular science literature dealing with epigenetic phenomena is rapidly increasing on a worldwide scale. Some examples of these books and articles are entitled *Identically Different: Why You Can Change Your Genes* (Spector, 2012), *The Genie in Your Genes: Epigenetic Medicine and the New Biology of Intention* (Church, 2009) and *Genes Are Not Destiny* (Blech, 2010; German title). A reoccurring motive in these works is the idea that, given full knowledge of the plasticity (and robustness) of epigenetic mechanisms, humans are able to unchain themselves from their genes as well as mould their organismic destiny and that of their ancestors. In other words, they renounce mono-causal and gene-centred or gene-essentialist determinations of mankind. A central issue is how to change, upgrade or improve your epigenome. This means enhancing or suppressing particular properties through induction, like a particular diet, physical activity, stress, alcohol, nicotine and pharmacological interventions.

The developmental and postgenomic perspective of the extended synthesis is taken both as starting point and justification for developing transhumanist, utopian proposals in which humans take a more active part in shaping their ontogenetic and phylogenetic future. Interestingly, this discourse does not emerge within the popular science literature or in the general public only, but is rather initiated through academic discussion in biology. In the current debate on expanding evolutionary theory, *biological concepts* like phenotypic plasticity, robustness (or canalization), organism, niche, natural selection and soft inheritance (also known as Lamarckian inheritance of acquired characteristics) are often mixed with *anthropological concepts* like identity or individuality, innateness, agency, autonomy or self-determination, destiny and responsibility. For example, evolutionary biologist and geneticist David Haig states:

> It seems to me that arguments between supporters and critics of the Modern Synthesis often are based on differences of preference and thinking style rather than matters of substance.... I will give my subjective impressions of the reasons why many people I have talked with, both in the general public and the scientific community, have a visceral attraction to Lamarckism and a visceral dislike of Darwinism.... A neo-Darwinian view, with its emphasis on chance and randomness in the origin of variation, is perceived as positing *a world without meaning* that is less attractive than a Lamarckian view in which organisms have *agency* in shaping their evolutionary *destiny*. *Natural selection*, with its reliance on differential survival and reproductive competition, is

> also perceived as bleak and harsh.... [I]t is a view from which many recoil. *Phenotypic plasticity* and the *inheritance of acquired characters* seem to hold hope that *we all can improve* without processes of selection.
>
> (Haig, 2011, p. 5, emphasis added)

Haig's description of the current postgenomic turn towards quasi-Lamarckian developmental processes in evolutionary biology is labelled by others as a turn towards a 'post-dichotomous view of *biosocial* processes' (Meloni, 2014b, p. 5, emphasis added) or, more generally, a '*social biology*' (Pickersgill *et al.*, 2013; Meloni, 2014a, 2014b). This new field investigates how the activity of the 'reactive genome' (Gilbert, 2003) is modulated through humans' actions and socio-economic environment, and vice versa. The social, anthropological and normative issues arising from this novel perspective, as well as their associated sociotechnical imaginaries, are grounded on two *bio-socio-anthropological conceptual clusters*. In cluster (i) the concepts of *biological organism*, *body*, and *identity*, and in cluster (ii) the concepts of *phenotypic plasticity*, *innateness*, *self-determination* and *responsibility* are commingled. Let me describe these clusters in more detail.

(i) Recently a number of biologists and philosophers of science have announced the 'return' of the organism (and its agency) in modern evolutionary biology (e.g. Bateson, 2005; Nicholson, 2014). Although such calls are not new, they have not been able to affect mainstream evolutionary theory since the establishment of the modern synthesis, which is first and foremost focused on the transmission of genes and its effect on populations, rather than on the developing organism. Later, in the 1960s to 1980s, the gene-centred view of evolution, defended famously by Richard Dawkins (1976), even reduced organisms to the rank of 'vehicles' of (egoistic) genes. According to this view, all evolutionary relevant processes are genetic processes. They program the development of organisms, or, as Dawkins calls them, 'survival machines' and 'lumbering robots'. Influenced by this development, the philosopher Michael Ruse (1989) even wrote a paper entitled 'Do organisms exist?'

Currently the traditional mechanistic conception of 'organism' is criticized. Biologists and philosophers emphasize that this view cannot make sense of the intrinsic purposefulness of organisms in contrast to the extrinsic purposefulness of machines (see e.g. Nicholson, 2013). Immanuel Kant (1913) has described this particular inner teleology as a result of the causal interdependencies between parts of organisms which are, in contrast to machines, self-organized beings. Within this conceptual framework, modern biologists seek to interlink evolutionary explanations of the phylogenetic teleology of organisms with developmental and molecular explanations of the purposeful organismic organization of self-organized and self-maintaining developmental systems. In other words, organismic teleology turns into an *explanans* in evolutionary theory. In addition, biological information is attributed to the organism itself, not just to parts of it, such as genes. It plays a crucial role in maintaining its own form of organization in a constantly changing environment.

Accordingly, this means that humans are *biologically* defined as the crucial authority, the decision maker, when it comes to realizing developmental potentials in the future. However, the burden of realizing these potentials is often not understood as being carried by single individuals alone. In contrast, the postgenomic and interactionist perspective, which focuses on the systemic relationships between developmental processes, purposeful actions, and environmental factors, suggests that adaptation and enhancement, respectively, are less egoistic and competitive processes of selection than cooperative processes taking place in the social sphere.

At this border between the biological and the social, Jörg Niewöhner (2011) describes a new concept of the *body* currently emerging in modern biology in general and epigenetics in particular – the so-called 'embedded body'. According to this concept, the human body is no longer a machine-like unit that is genetically programmed, neurally controlled and bounded by the skin but an open and dynamic unit that cannot be grasped in isolation from its material and social environment. Additionally, the body is embedded into different time scales ranging from its evolutionary and transgenerational to ontogenetic past, which permanently constitute its present.

How does this systemic concept of organism and spatially and temporally open concept of the body highlighted by epigenetics or postgenomics affect the notion of (biological) *identity* of living beings in general and humans in particular? In other words, if gene-essentialism is wrong and if we are not our genes, as it is commonly assumed nowadays, what are we humans instead?

Giovanni Boniolo and Giuseppe Testa (2012) argue that epigenetics helps us solve this problem by answering two questions: first, '*who* is that living being?' (i.e. what makes it unique), and, second, 'how does it persist from one time to another?' (i.e. what makes it remain the same over time, given that certain of its properties change). The latter issue of diachronic identity, for example, refers to the problem that an embryo and adult organism can be the same living being at different points in time. Most imaginaries of enhanced organisms also deal with this issue, since enhancement should usually preserve identity.

According to Boniolo and Testa, epigenetics teaches us that, in order to solve these problems of identity we need to focus not on single properties, like phenotypic (or mental) properties, but on the whole phenotype, which is developed in an environmentally sensitive process of epigenetic gene regulation and maintained over time as a robust developmental system.[3] Thus, for example, diachronic identity can be understood as the continuity of the whole phenotype (not of its parts) over time, guaranteed by the continuity of epigenetic processes. In this case, these processes allow for the robustness of the organism in interaction with the environment and thus for the persistence of identity over time.

In short, the continuity of individuals' epigenetic history constitutes the identity of single- and multicellular organisms. Note that, given the phenomenon of *transgenerational* continuity of epigenetic processes, Boniolo's and Testa's view is radical. If in the following generation(s) a previous phenotype is epigenetically maintained, their view implies that living beings can no longer be understood as (at least not qualitatively) delimitated over time by mitotic or meiotic cycles.

More generally, these discourses on the *'openness' of the body* and *generation-overarching identity* show that in recent epigenetic imaginaries of the future, human enhancement is understood as a systemic process in which individuals living in different temporal zones reciprocally interact and the ontogenetic autonomy of living beings takes a back seat.

(ii) Epigenetics suggests a non-teleological understanding of development as a process starting from a particular genotype, not directed towards a particular phenotype. As described above, this openness of epigenetically regulated developmental processes is commonly labelled *phenotypic plasticity*. The degree of stability or malleability of environmentally induced and epigenetically modulated variation fluctuates on a case-by-case basis. Epigenetic markers are described by biologists as 'potentially reversible', yet also as 'long-lasting', thus ranging from epigenetic plasticity to epigenetic determinism.[4]

In popular science literature, epigenetics is often allocated the salutary utopian function of rescuing humankind from its genes. It is understood as a corrective which might take two different forms. The first is the form of a *determining corrective*. This is the case, if, for example, parental decision-making affects the development of their children (and further generations) irreversibly or induces the cyclic process described above, in which a particular socio-economic 'niche' is re-established in every subsequent generation. This motive is expressed, for example, in newspaper headlines such as 'Does dad's diet determine a baby's genetic fate?' or 'Babies born into poverty are damaged forever before birth' (McLaughlin, 2012; Mohan, 2013). In this case, an embryo's innate properties and developmental destiny are no longer contingent on genetic determinants but environmental determinants of its epigenetic linage.

Often, however, the epigenetic corrective is understood by means of the idea of *plastic destiny*. This means that by emphasizing the reversibility of epigenetic markers a new view of development and inheritance is adopted, one that is incompatible with the previous hierarchical mono-causal concept of innateness – a concept which is tightly linked to gene-essentialism and the assumption that a temporally and informationally privileged ontological unit exists in living systems (see Meloni, 2013, p. 94ff.). In other words, the prevailing view of the gene as an autonomous and stable original unit is replaced by a postgenomic view of contextualized genes and non-linear developmental dynamics in which there simply are no original privileges.

In these dynamics, humans play a special role as causal agents, because they have (in the best case, full) knowledge about the molecular and developmental processes underlying their plasticity. In other words, knowledge allows humans to create their ontogenetic and phylogenetic future in a *self-determined* manner. This implies the possibility to enhance skills and talents, as Maurizio Meloni (2014b, p. 7) notes: '[T]he upgradable epigenome may become the basis for a new motivation to intervene, control and improve it.' Although such an enhancement can be conducted therapeutically or pharmacologically, often a bodily change is advised that proceeds as a non-technologically mediated modification of everyday behaviour.

These utopian imaginaries of epigenetic enhancement directly lead to the question 'who is *responsible* for such actions of improvement?' You should notice, first, that we cannot directly derive from knowledge about epigenetic regulatory processes normative prescriptions like 'health must be facilitated, welfare improved and violence and stress avoided' (see Schuol, 2014). This naturalistic fallacy is sometimes even committed by philosophers and social scientists who argue that the new social biology, including epigenetics, contributes to 'moralising biology' (Meloni, 2013). However, factual knowledge in epigenetics does not per se imply fulfilling norms or being responsible. It always needs additional normative premises to avoid a naturalistic fallacy. In fact, such a premise could be the rule 'health must be facilitated, welfare improved and violence and stress avoided' itself.

Who, however, takes responsibility for this rule? If epigenetics is not committed to fulfilling it, who is? Is it *personal responsibility* that matters in this case, and, if it does, how can we characterize it in more detail? As Niklas Luhmann (1990) emphasized, when it comes to attributing responsibility for certain events we have to distinguish between danger (*Gefahr*) and risk (*Risiko*). In the first case, avoidable damage is brought about by a natural force or 'environmental decisions', whereas in the latter the damage is brought about by human decisions or actions. Against this background, Sebastian Schuol (2014) asks whether epigenetics introduces a new kind of personal responsibility towards avoiding diseases. While diseases have previously been a danger, they now seem to become a risk – a personally avoidable damage.[5] This means that individuals are now responsible for whether or not they suffer, for example, from obesity and cancer. The high emotional stress resulting from this responsibility is even reinforced if we consider the transgenerational responsibilities of individual 'care-takers of life that does not yet exist' (Pickersgill *et al.*, 2013, p. 437).

Previously, parents have very rarely been deemed guilty for passing down to their children a disposition to a particular disease. In contrast, given (future) possibilities of preventing epigenetically inherited diseases, parents and – due to their various ways of directly affecting epigenetic inheritance pathways during pregnancy – mothers in particular may now be declared guilty in a number of new ways.

In face of the high emotional stress and the merely stochastic effects of human interventions on the complex and non-linear processes of epigenetic developmental systems, bioethicists and political scientists have recently come out against the concepts of epigenetic personal responsibility. In contrast, it is expanded towards a web of *collective responsibility* (see Hedlund, 2012; Dupras, Ravitsky and Williams-Jones, 2014). In line with the interactionist, postgenomic perspective of epigenetics, according to which developmental systems are constituted by multi-factorial, non-linear processes instead of linear processes, humans also have to be understood as – to use Niewöhner's terminology – 'embedded' in complex causal networks. Only an individual who is part of a community can affect and enhance his/her environment and himself/herself as well as take responsibility for these actions. Thus, it is first and foremost the sociopolitical, governmental level of states and the international field where responsibility for actions of epigenetic enhancement should be taken. Here, material and social resources should be provided for individuals – the citizens – to mould their social, economic and cultural 'niche' in such a way that transgenerational vicious circles can be avoided and moral rules fulfilled. Finally, this may lead to an 'upward spiral' towards a new and hopefully better society.

The emergence of two new conceptual clusters described above may be understood as the most recent development in the long history of interrelating biological and social as well as anthropological concepts. For example, not least since the rise of genetic technologies, the idea of social identity has been influenced by scientists' construction of biological identities. As Rabinow (1992) emphasizes, knowledge about, for example, genetic disease susceptibility through diagnostic procedures has influenced the formation of individuals' identity and thus social relations. He calls these newly formed social identities 'biosociality'. This interrelatedness not only facilitates shared group identity and collectiveness but also public activism, as 'genetically formed' social groups try to change their biological identity – often for the better. In a similar manner, Fox-Keller (2002) points to the intertwining of explanations and concepts in the biosciences with social and political standpoints. This connectedness makes it impossible to understand nature outside of a socially constructed arena. For example, she argues that the concept of masculinity, prominent in the public sphere, has widely affected how population geneticists have modelled evolutionary change, i.e. by omitting sexual difference and reproduction (Fox-Keller, 1992).

Although standing in the tradition of this bio-social-interplay, the more recent two *bio-socio-anthropological conceptual clusters* that arise in current academic and public discourse on epigenetics go along with unique sociotechnical imaginaries of humans' future and of how to enhance human bodies and societies. Despite being quite different, these imaginaries should nevertheless be conceived as interlinked rather than strictly distinct. Cluster (i) described above is coupled with a view that highlights the 'openness' and 'embeddedness' of the body in past, present and future societies in which individuals show transgenerational identity. The other conceptual cluster, (ii), underlies two opposing imaginaries of a future society: one imaginary that guarantees individuals 'plastic' destiny and self-determination but also gives them various responsibilities, especially individual accountability for health outcomes; and another one that, due to epigenetic or environmental determination on the level of individual agents, ascribes the possibilities and responsibilities of changing ontogenetic and phylogenetic trajectories to governmental or international (i.e. collective) policymakers instead of to individuals. These different imaginaries affect science and society in a growing number of ways.

Future directions in science and public health

As described above, the conceptual framework of a new theory of 'biosocial processes', or, in short, a new social biology, is used as a starting point to develop novel perspectives on how to transform human bodies and societies. This development has two wide-ranging consequences: First, it affects scientific research by redirecting funding streams in the biosciences and legitimizing new research projects. Second, it stirs a variety of new health policy programmes in which individual and/or social enhancement is realized through a number of policy-makers, ranging from individuals to collectives and from governmental to non-governmental agents. These health applications are still immature, however, as they are developed out of contradicting views of future societal responsibilities. Let us discuss these two developments in more detail.

First, epigenetic imaginaries have contributed to stirring and legitimizing particular funding streams in the biosciences. In the last ten years, funding in epigenetics has increasingly shifted to biomedical sciences and applied behavioural epigenetics, which includes oncology in particular but also research on diseases such as ageing, arthritis, Alzheimer's disease, anxiety and risk-taking, stress, learning and memory, mental health, eating disorders and obesity, schizophrenia, bipolar disorder, major depressive disorder and autism. Not least owing to the fact that imaginaries of possible epigenetic enhancement and plastic bodies have recently been spreading in academic discourse and, especially, among the public, research in the diagnostic segment and on therapeutic or pharmacological modifications started to boom. According to a recent market report, the value of the global market for epigenetic drugs and diagnostic technologies is estimated to reach USD 5.7 billion in 2018 and the market of Histone Deacetylase Inhibitors (HDAC) alone is estimated to grow each year by as much as 30 per cent (from 2013–2020: Allied Market Research, 2015). Of course, this development has been accompanied by a rapid increase in market-driven research, as pharmaceutical companies began to run their own epigenetic research centres (e.g. Novartis, based in Switzerland, which has more than 200 employees working in epigenetics, especially on cancer).

There is another important development in the biosciences. Driven not least by growing public concerns about the low causal responsibility of single individuals' actions on some complex developmental and inheritance phenomena such as cancer, epigenetic changes are more and more investigated in a systemic, genome-wide manner. These epigenomic studies, run in labs of pharmaceutical companies and in programmes funded by the European Union or the NIH (US National Institutes of Health), such as the BLUEPRINT project or the Roadmap Epigenomics Project, produce huge volumes of data which are hard to make sense of. As a consequence, the field desperately needs talented informaticians (Hayden, 2012). This new trend in epigenetics' job market and hiring strategies involves the danger that the field's direction and, as a consequence, the public picture of epigenetic processes, is increasingly drawn by scientists trained in computation but not in biology. Time will tell how this will influence our understanding of living phenomena in the future.

More generally, imaginaries of humans' bodily interwovenness with their future have provoked and legitimized the study of a variety of diseases from both intra- and intergenerational perspectives. For example, Marcus Pembrey, professor of paediatric genetics at University College London, emphasizes: 'I suspect we will not understand the rise in neuropsychiatric disorders or obesity, diabetes and metabolic disruptions generally without taking a multigenerational approach' (Sergo, 2013). However, this new perspective on public health involves focusing not only on new determining (non-genetic) factors, but on new pathways of inheritance, new risks and/or dangers, new challenges, new responsibilities and, more generally, new favourable ways of living and desirable social structures.

The latter directly leads to a second major effect of epigenetic imaginaries: novel societal policymaking. This development, however, is still in its infancy and highly heterogeneous. Based on different views of who should be accountable for creating a healthier future society, it includes applications that address either individuals as health and well-being agents or economic, governmental and non-governmental collectives, often in a contradicting manner.

As a result of the public hopes and fears that are stimulated by reports on the new plasticity and malleability of humans' (future) bodies, individuals are asked to take better care of themselves and their children and grandchildren. For example, advice literature that offers prescriptive instructions for individuals' life praxis, especially for pregnant women, is in full bloom. For example, Johannes Huber (2010) describes in his book *Love Can be Inherited* (German title) what he calls an 'epigenetic of stroking', which should biologically strengthen the mother–child relationship.

As Richardson *et al.* (2014) argue, this view is deeply intertwined with research in the field called DOHaD (Developmental Origin of Health and Disease). Based on investigations on how epigenetic factors modulate environmental input, DOHaD should ideally guide policies that support parents and children; however, in this field, 'exaggerations and over-simplifications are making scapegoats of mothers, and could even increase surveillance and regulation of pregnant women' (2014, p. 131). According to Richardson and colleagues, DOHaD research ties in with a long history of societal blaming of mothers for their children's health, such as media hysteria and governmental overreaction on 'refrigerator mothers' lacking emotional warmth or mothers consuming alcohol during pregnancy in the 1970s and 1980s. One of the things they emphasize is that many intrauterine stressors have intergenerational effects that can only be influenced at the level of the society.

Unfortunately, these ideas of individual malleability, responsibility and guilt have already begun influencing governmental policy programmes. Currently a number of consultants in public health inform health officials on the state and international level about epigenetic research. For example, at a symposium for Public Health Wales (held in July 2014 at the Temple of Peace and Health in Cardiff for employees and a number of Welsh government policymakers), Layla Jader, consultant in public health genomics, called for a new policy built on epigenetic research results that would target both the health and education of children. She argued for developing an integrated, comprehensive and sustained long-term strategy that includes drilling the message into women: 'Look after yourself; you are the human incubator for your children and your grandchildren' (Jader, 2014).

At the same time and somewhat contradictory to the above, Jader releases the individuals, such as mothers and children, from being the prime health care agents, insofar as she also asks the Welsh government to take charge of developing health and education applications. This includes offering better incentives for the best-trained students to become public health agents, such as midwives, school nurses and teachers in antenatal clinics, preschools and schools. As a consequence, teaching and prevention become deeply intertwined.[6] Moreover, in line with those social and cultural scientists, arguing for a concept of collective responsibility for epigenetic modifications, Jader calls for reinforcing employment laws with respect to maternity as well as child care and protection. This and other, similarly 'society-orientated' advice seems to adopt ideas of the growing paradigm of 'environmental justice' (Schlosberg, 2007), which emphasizes the fair distribution of environmental benefits and burdens controlled by environmental law and governance, as well as environmental planning and policy.

The underlying imaginary of developing a healthier future society collectively, rather than everyone for himself or herself, has recently become very popular not only in governmental circles but in the economy, such as in the food industry. For example, Nestlé donates CHF 22 million to a six-year research partnership with an international alliance of epigenetic researchers. They do so, as Heiko Schipper, CEO of Nestlé Nutrition, argues, because Nestlé understands itself as a 'Nutrition, Health and Wellness company' that is responsible for 'improving the quality of peoples' lives' (Nestlé, 2014). It is hard to tell whether the company's slogan to 'nurture a healthier generation' arises in fact from collective responsibility or is just a clever advertising pitch, because the idea of feeding embedded bodies means not only caring for multiple generations at once but also attracting more consumers.

As these brief remarks show, currently a number of different and even conflicting epigenetic imaginaries of the future have begun influencing policymaking on various societal levels. We find more and more policy programmes in which individuals are asked for the establishment of a *society of caring for oneself* that includes the idea of personal accountability for health outcomes. Other – or sometimes even the very same – programmes call for a *society of caring for others*, in which groups and collectives in particular are asked to act according to the principles of relational care, and co-responsibility.

The 'new' social and future challenges

Of course, the concept of the social has been around in biology for quite some time, from Darwin's work on sexual selection to the concept of niche in ecology.

Most prominently in Dawkins's sociobiology many social phenomena were considered as epiphenomena of the 'actions' of selfish genes – i.e. merely as extended phenotypes (Dawkins, 1982). Julian Huxley, best known for his seminal work *Evolution: The Modern Synthesis* (1942), had a similarly gene-centred view of the social. In 1939, together with a number of renowned biologists, he wrote the so-called 'geneticists' manifesto', which carried the title 'Social Biology and Population Improvement'. This liberal eugenic programme aimed at answering the question 'How could the world's population be improved most effectively genetically?' (Crew *et al.*, 1939, p. 521). The answer was: By modifying genetic transmission through new technologies of reproductive medicine and manipulating gene frequencies in populations.

Rather than treating social life as (in large part) genetically determined (Dawkins), or as improvable through a eugenic programme that focuses on populations' gene frequencies (Huxley), the recent postgenomic turn in general and epigenetics in particular has triggered a number of new imaginaries of future social life. These epigenetic imaginaries emerge from a new link between concepts from modern molecular and developmental biology, such as phenotypic plasticity, organism, etc., with anthropological, social and normative concepts, like body, identity, innateness, self-determination and responsibility. In more general terms, what we learn from this recent epigenetic turn is how sociotechnical imaginaries are co-constructed conceptually in the biosciences and the public. Moreover, these new epigenetic imaginaries have begun demonstrating and targeting the different paths humans should take to manifest a better society. At the current state, these different paths are not accompanied by clear-cut and full-blown policy agendas. They are, rather, different social-utopian views of individually, governmentally and/or technologically guided social enhancement that might, in fact, fall foul of each other, especially when it comes to questions of accountability, responsibility and guilt.

As a result, there are various issues to be addressed in future investigations of this new social biology and its societal consequences. First, although the current epigenetic turn is usually understood to be a worldwide movement, it is useful to strengthen national and cross-national investigation. For example, advice for policymaking often includes the idea of, as Layla Jader (2014) puts it, 'defining the long-term development path for our nation' in order to build sustainable countries with 'healthy, productive people'. Thus, similar to recent studies focusing on ways of imagining improvement of health and society in the 'genetic age' (Jasanoff, 2011), future national case studies should look at the construction and societal impact of new imaginaries in the *postgenomic* or *epigenetic age*. An interesting study could be whether a nation-specific increase in ageing population and cancer leads to different imaginaries and thus, for example, different investments in research, epigenetic drugs, diagnostic technologies and health care policies. For example, it is currently expected that because Asia accounts for nearly half of the new cancer cases worldwide, this demand will heavily affect the market of epigenetic drugs and diagnostic technologies in the region. Another crucial question is how, from a national or cross-national perspective, responsibility for actions of epigenetic modification is distributed within nations and across generations. Germany's 'epigenetic turn' would be a very interesting case, because of the nation's long history in debating individual vs. collective and intragenerational vs. transgenerational guilt.

Second, it should be investigated how this recent marriage between the social and the biological emerging in the age of epigenetics in detail differs from those previous connections between the two realms described earlier by Rabinow (1992), Fox-Keller (1992, 2002) and others. For example, with respect to Rabinow's concept of 'biosociality' one clearly sees that novel 'epigenetic collectivities' are no longer solely formed 'at' the molecular or even genetic level only but rather the organismic level as well as through organism–environment interactions. As a consequence, the recent postgenomic perspective, in principle, allows for a non-technological (but behaviourally) mediated construction of one's epigenetic and thus social identity. This opens up a whole new dimension of empowerment of groups and individuals, which may facilitate strong public activism promoting epigenetic research. What is more, epigenetics and postgenomics more generally hold the promise of a biosociality in which the contribution of the biological is at least in principle detached from the paradigm of biological determinism, as well as decoupled from the previously dominant idea of masculinity, criticized by Fox-Keller.

Finally, and related to the above, one has to investigate whether we are currently witnessing a new era of a *'socialization' of the biological world* or a *'biologization' of the social world.* On the one hand, owing to the new social biology the developmental niches of organisms seem to turn into social phenomena that are approached with methodologies and modelling tools of the social sciences. On the other hand, this development also includes, for example, a molecularization of the biography and the social milieu of humans. This means that Pierre Bourdieu's (1979) economically, institutionally and socioculturally defined concept of the milieu is significantly expanded. Now, the milieu literally gets under the skin of the social agent. It influences his/her molecularly describable 'habitus', which can be passed on through generations *biologically*. Although this new trend towards biologization does not per se imply gene- or epigene-deterministic imaginaries of humans' future, social phenomena are looked at through biology-coloured lenses. Time will tell in which of these two directions epigenetic research associated with novel ideas and imaginings of scientific progress and social order will develop.

Notes

1 Acknowledgements: I thank the participants of the 'Science Shaping the World of Tomorrow' workshop (Antwerp, 2015), as well as the editors, for constructive comments on earlier versions of this chapter. Financial support from the German Research Foundation (DFG; project no. BA 5808/1–1) is gratefully acknowledged.

2 Evolvability is the capacity of a system to generate heritable, adaptive phenotypic variation and thus to evolve in evolution (see Hendrikse, Parsons and Hallgrímsson, 2007).

3 This robustness includes, for example, that after its epigenetically 'programmed' development from a stem cell a liver cell functionally stays the same over many cell divisions and changes in environmental factors.

4 On epigenetic determinism, see also Waggoner and Uller (2015).

5 In fact, Luhmann already notes that this trend holds for a couple of other natural disasters, such as floods, which are nowadays considered as 'risks' due to a lack of humans' preventive measures.

6 For an example of a national prevention programme at schools, see the 'Teens4Kids – Gesundheit macht Schule!' ('Health becomes the thing!') programme, developed by a collective of German health foundations and the Frankfurter Allgemeine Zeitung (FAZ); see Teens4Kids (2015).

References

Allied Market Research, 2015. *Histone deacetylase inhibitors market: opportunities and forecasts, 2014–2020, allied market research report*. Available at: www.alliedmarketresearch.com/histone-deacetylase-inhibitors-market [Accessed 30 September 2015].

Anway, M., Cupp, A., Uzumcu, M. and Skinner, M., 2005. Epigenetic transgenerational actions of endocrine disruptors and male fertility. *Science*, 308(5727), pp. 1466–1469.

Arai, J., Li, S., Hartley, D. and Feig, L., 2009. Transgenerational rescue of a genetic defect in long-term potentiation and memory formation by juvenile enrichment. *The Journal of Neuroscience*, 29(5), pp. 1496–1502.

Arrowsmith, C., Bountra, C., Fish, P., Lee, K. and Schapira, M., 2012. Epigenetic protein families: a new frontier for drug discovery. *Nature Reviews Drug Discovery*, 11(5), pp. 384–400.

Bateson, P., 2005. The return of the whole organism. *Journal of Biosciences*, 30(1), pp. 31–39.

Blech, J., 2010. *Gene sind kein Schicksal: Wie wir unsere Erbanlagen und unser Leben steuern können*. Frankfurt am Main: Fischer.

Boniolo, G. and Testa, G., 2012. The identity of living beings, epigenetics, and the modesty of philosophy. *Erkenntnis*, 76(2), pp. 279–298.

Borghol, N., Suderman, M., McArdle, W., Racine, A. Hallett, M., Pembrey, M., Hertzman, C., Power, C. and Szyf, M., 2012. Associations with early-life socio-economic position in adult DNA methylation. *International Journal of Epidemiology*, 41(1), pp. 62–74.

Bourdieu, P., 1979. *La distinction: Critique sociale du jugement*. Paris: Minuit.

Church, D., 2009. *The genie in your genes: epigenetic medicine and the new biology of intention*. Santa Rosa: Elite.

Crew, F.A.E., Darlington, C.D., Haldane, J.B.S., Harland, S.C., Hogben, L.T., Huxley, J.S., Muller, H.J., Needham, J., Child, G.P., David, P.R., Dahlberg, G., Dobzhansky, T., Emerson, R.A., Gordon, C., Hammond, J., Huskins, C.L., Koller, P.C., Landauer, W., Plough, H.H., Price, B., Schultz, J., Steinberg, G. and Waddington, C.H., 1939. Social biology and population improvement. *Nature*, 144(3646), pp. 521–522.

Crick, F., 1958. On protein synthesis. *Symposium of the society of experimental biology*, 12, pp. 138–163.

Dawkins, R., 1976. *The selfish gene*. Oxford: Oxford University Press.

Dawkins, R., 1982. *The extended phenotype*. London: Freeman.

Dunn, G. and Bale, T., 2011. Maternal high-fat diet effects on third-generation female body size via the paternal lineage. *Endocrinology*, 152(6), pp. 2228–2236.

Dupras, C., Ravitsky, V. and Williams-Jones, B., 2014. Epigenetics and the environment in bioethics. *Bioethics*, 28(7), pp. 327–334.

Fox-Keller, E., 1992. *Secrets of life, secrets of death*. New York, NY: Routledge.

Fox-Keller, E., 2002. *Making sense of life: explaining biological development with models, metaphors, and machines*. Cambridge, MA: Harvard University Press.

Gilbert, S., 2003. The reactive genome. In: G.B. Müller and S.A. Newman, eds. 2008. *Origination of organismal form: Beyond the gene in developmental and evolutionary biology*. Cambridge, MA: MIT Press. pp. 87–101.

Haig, D., 2011. Lamarck ascending! *Philosophy & Theory in Biology*, 3(e204), pp. 1–6.

Hayden, E., 2012. Biostatistics: revealing analysis. *Nature*, 482(7384), pp. 263–265.

Hedlund, M., 2012. Epigenetic responsibility. *Medicine Studies*, 3, pp. 171–183.

Hendrikse, J.L., Parsons, T.E. and Hallgrímsson, B., 2007. Evolvability as the proper focus of evolutionary developmental biology. *Evolution & Development*, 9(4), pp. 393–401.

Huber, J., 2010. *Liebe lässt sich vererben: Wie wir durch unseren Lebenswandel die Gene beeinflussen können*. Munich: Zabert Sandmann.

Huxley, J., 1942. *Evolution: the modern synthesis*. London: Allen & Unwin.

Jablonka, E., 2006. Genes as followers in evolution – a post-synthesis synthesis? *Biology & Philosophy*, 21(1), pp. 143–154.

Jablonka, E. and Lamb, M., 1995. *Epigenetic inheritance and evolution: the Lamarckian dimension*. Oxford: Oxford University Press.

Jablonka, E. and Lamb, M., 2005. *Evolution in four dimensions: genetic, epigenetic, behavioral, and symbolic variation in the history of life*. Cambridge: MIT Press.

Jablonka, E. and Lamb, M., 2010. Transgeneration epigenetic inheritance. In: M. Pigliucci and G.B. Müller, eds. 2010. *Evolution: the extended synthesis*. Cambridge: MIT Press. pp. 137–174.

Jader, L., 2014. Epigenetics symposium. In: *Public Health Wales*. Cardiff, UK, 10 July 2014. Available at: www.wales.nhs.uk/sitesplus/888/page/75795 [accessed 30 September 2015].

Janzen, F. and Phillips, P., 2006. Exploring the evolution of environmental sex determination, especially in reptiles. *Journal of Evolutionary Biology*, 19(6), pp. 1775–1784.

Jasanoff, D. and Kim, S.-H., 2009. Containing the atom: sociotechnical imaginaries and nuclear power in the United States and South Korea. *Minerva*, 47(2), pp. 119–146.

Jasanoff, S., 2011. *Reframing rights: bioconstitutionalism in the genetic age*. Cambridge, MA: MIT Press.

Kaati, G., Bygren, L. and Edvinsson, S., 2002. Cardiovascular and diabetes mortality determined by nutrition during parents' and grandparents' slow growth period. *European Journal of Human Genetics*, 10(11), pp. 682–688.

Kant, I., 1913. *Kritik der Urteilskraft* [*Kant's gesammelte Schriften V*, edited by Königliche Preußische Akademie der Wissenschaften]. Berlin: Reimer. [Original: 1790/93]. pp. 165–485.

Kappeler, L. and Meaney, M., 2010. Epigenetics and parental effects. *BioEssays*, 32(9), pp. 818–827.

Laland, K., Uller, T., Feldman, M., Sterelny, K., Müller, G., Moczek, A., Jablonka, E. and Odling-Smee, J., 2015. The extended evolutionary synthesis: its structure, assumptions and predictions. *Proceedings of the Royal Society B*, 282(1813), p. 20151019. doi:10.1098/rspb.2015.1019.

Luhmann, N., 1990. Risiko und Gefahr. In: N. Luhmann, *Konstruktivistische Perspektiven*. Opladen: Westdeutscher Verlag. pp. 131–169.

McLaughlin, M., 2012. Babies born into poverty are damaged forever before birth. *The Scotsman*, 24 January. Available at: www.scotsman.com/news/scotland/top-stories/babies-born-into-poverty-are-damaged-forever-before-birth-1-2072713 [accessed 30 September 2015].

Meloni, M., 2013. Moralizing biology: the appeal and limits of the new compassionate view of nature. *History of the Human Sciences*, 26(3), pp. 82–106.

Meloni, M., 2014a. How biology became social, and what it means for social theory. *The Sociological Review*, 62(3), pp. 593–614.

Meloni, M., 2014b. The social brain meets the reactive genome: neuroscience, epigenetics and the new social biology. *Frontiers in Human Neuroscience*, 8(309), pp. 1–12.

Minelli, A., 2009. *Forms of becoming: the evolutionary biology of development*. Princeton, NJ: Princeton University Press.

Mohan, G., 2013. Does dad's diet determine a baby's genetic fate? *LA Times*, 11 December. Available at: www.latimes.com/science/sciencenow/la-sci-sn-epigenetics-folate-fathers-20131210-story.html [accessed 30 September 2015].

Nestlé, 2014. *Nestlé boosts research into cutting-edge maternal nutrition and epigenetics*. Available at: www.nestle.com/media/newsandfeatures/nestle-research-epigenetics [accessed 30 September 2015].

Nicholson, D., 2013. Organisms ≠ machines. *Studies in History and Philosophy of Biological and Biomedical Sciences*, 44(4), pp. 669–678.

Nicholson, D., 2014. The return of the organism as a fundamental explanatory concept in biology. *Philosophy Compass*, 9(5), pp. 347–359.

Niewöhner, J., 2011. Epigenetics: embedded bodies and the molecularisation of biography and milieu. *BioSocieties*, 6(3), pp. 279–298.

Odling-Smee, J., 2010. Niche inheritance. In: M. Pigliucci and G. Müller, eds. 2010. *Evolution: The extended synthesis*. Cambridge, MA: MIT Press. pp. 175–207.

Odling-Smee, J., Laland, K. and Feldman, M., 2003. *Niche construction: the neglected process in evolution*. Princeton, NJ: Princeton University Press.

Painter, R., Osmond, C., Gluckman, P., Hanson, M., Phillips, D. and Roseboom, T., 2008. Transgenerational effects of prenatal exposure to the Dutch famine on neonatal adiposity and health in later life. *British Journal of Obstetrics and Gynaecology*, 115(10), pp. 1243–1249.

Pembrey, M., Bygren. L., Kaati, G., Edvinsson, S., Northstone, K., Sjöström, M., Golding, J. and ALSPAC Study Team, 2006. Sex-specific, male-line transgenerational responses in humans. *European Journal of Human Genetics*, 14(2), pp. 159–166.

Pickersgill, M., 2011. Connecting neuroscience and law: anticipatory discourse and the role of sociotechnical imaginaries. *New Genetics and Society*, 30(1), pp. 27–40.

Pickersgill, M., Niewöhner, J., Müller, R., Martin, P. and Cunningham-Burleya, S., 2013. Mapping the new molecular landscape: social dimensions of epigenetics. *New Genetics and Society*, 32(4), pp. 429–447.

Pigliucci, M. and Müller, G., 2010. *Evolution: the extended synthesis*. Cambridge, MA: MIT Press.

Plagemann, A., 2012. Towards a unifying concept on perinatal programming: vegetative imprinting by environment-dependent biocybernetogenesis. In: A. Plagemann, ed. 2012. *Perinatal programming: the state of the art*. Berlin: de Gruyter. pp. 243–282.

Rabinow, P., 1992. Artificiality and enlightenment: From sociobiology to biosociality. In: J. Crary and S. Kwinter, eds. 1992. *Incorporations*. New York, NY: Zone. pp. 234–252.

Radtke, K., Ruf, M., Gunter, H., Dohrmann, K., Schauer, M., Meyer, A. and Elbert, T., 2011. Transgenerational impact of intimate partner violence on methylation in the promoter of the glucocorticoid receptor. *Translational Psychiatry*, 1(e21), pp. 1–6.

Richardson, S., Daniels, C., Gillman, M., Golden, J., Kukla, R., Kuzawa, C. and Rich-Edwards, J., 2014. Society: don't blame the mothers. *Nature*, 512(7513), pp. 311–312.
Rodriguez-Paredes, M. and Esteller, M., 2011. Cancer epigenetics reaches mainstream oncology. *Nature Medicine*, 17(3), pp. 330–339.
Ruse, M., 1989. Do organisms exist? *American Zoologist*, 29(3), pp. 1061–1066.
Sansom, R. and Brandon, R., 2007. *Integrating evolution and development*. Cambridge, MA: MIT Press.
Schlosberg, D., 2007. *Defining environmental justice: theories, movements, and nature*. Oxford: Oxford University Press.
Schuol, S., 2014. Kritik der Eigenverantwortung: Die Epigenetik im öffentlichen Präventionsdiskurs zum metabolischen Syndrom. In: V. Lux and J.T. Richter, eds. 2014. *Kulturen der Epigenetik: vererbt, codiert, übertragen*. Berlin: de Gruyter. pp. 271–281.
Sergo, P., 2013. The family tree of phobia: epigenetics explains how we inherit fear from our ancestors. *Medical Daily*, 2 December. Available at: www.medicaldaily.com/family-tree-phobia-epigenetics-explains-how-we-inherit-fear-our-ancestors-264134 [accessed 30 September 2015].
Spector, T., 2012. *Identically different: why you can change your genes*. London: Weidelfeld & Nicolson.
Stotz, K., 2008. The ingredients for a postgenomic synthesis of nature and nurture. *Philosophical Psychology*, 21(3), pp. 359–381.
Teens4Kids, 2015. *Epigenetik als Impuls für die Primärprävention des Metabolischen Syndroms – Das Konzept der Vegetativen Prägung*. Available at: www.teens4kids.de/epigenetik-als-impuls-fuer-die-primaerpraevention-des-metabolischen-syndroms-das-konzept-der-vegetativen-praegung [accessed 30 September 2015].
Waddington, C., 1952. *The epigenetics of birds*. Cambridge: Cambridge University Press.
Waddington, C., 1975. Introduction. In: Ciba Foundation, ed. 1975. *The future as an academic discipline*. Amsterdam: Elsevier. pp. 1–4.
Waggoner, M. and Uller, T., 2015. Epigenetic determinism in science and society. *New Genetics and Society*, 34(2), pp. 177–195.

4 Working imagination along the food–drug divide

Kim Hendrickx

Introduction

Since the 1990s, food producers have increasingly been making health claims about various packaged products, ranging from yoghurts and margarines to biscuits and soda drinks. National authorities have since drafted legislation to limit the scope of possible claims and to assess the scientific validity of claims. In the EU, a regulation, Regulation EC no. 1924/2006, entered into force in 2007. This was an important event, as it turned the ill-defined and debated 'health claim' into an *object of government*, thereby defining its content and scope. In this chapter I focus on one particular source that has provided key concepts for the EU regulation on health claims: an industry-funded think tank and its consortium of life scientists: the European branch of the International Life Sciences Institute (ILSI Europe, hereafter 'ILSI'). As a think tank, its job is to 'imagine' and – more precisely – to *formulate problems* with respect to their agenda, and to *anticipate* future challenges with respect to science, politics and regulation in the EU. ILSI's particular concepts have been influential in the drafting of the EU regulation, and these concepts are mobilized by scientists and food companies whenever a health claim is questioned or rejected by the European Food Safety Authority, which is in charge of the pre-market approval procedure for health claims on the EU market. I have detailed the problems and debates between industry and EFSA after the implementation of the regulation elsewhere (Hendrickx, 2013, 2014). What I would like to focus upon here is the historical emergence of certain concepts, some of which have been taken up in European legislation and that are still being defended today – most notably by adherents of *evidence-based nutrition* – in order to demarcate food from drugs as objects and as markets.

The theme of this book is the relationship between science and collective imaginaries in shaping the future. These are interesting concepts and heuristics, yet they can mean very different things. I will argue that, on the border between food and drugs, 'science' is rhetorically mobilized to refer to particular concepts and technologies of demonstration. Certain food ingredients are imagined to be good for one's health, but how does one demonstrate that? And how does one prevent this very demonstration from being taken as a token of the therapeutic capacity of food ingredients, thereby qualifying them as drugs? Avoiding this tipping point is the main stake in the health claims debate and ILSI's work in this area. I will show how ILSI engages in defining health differently and imagining the body differently than in clinical trials for drug testing.

ILSI's main concepts and strategies for health claims and 'functional foods' were formulated in two widely cited publications from the ILSI consortium. Both publications are position papers that were the outcome of projects coordinated by ILSI and financed under the European Commission's 4th and 5th Framework Programmes for innovative research. I propose a close reading of both papers, because they attest to the imagination of the scientists involved indeed, and more particularly because the analysis will allow to characterize the nature of that 'imagination': the translation of legal, political and commercial constraints into nutritional concepts. These constraints can be summed up as follows: *legally*, no genuine therapeutic properties may be attributed to food; *politically*, human health is related to a sectorial divide between food and drugs; and, *commercially*, the food industry is not interested in the strictly regulated market of pharmaceutical agents. Throughout the analysis, it will become clear that a particular understanding of the human body is co-produced with the types of trials and evidence for health claims that respond to the constraints just highlighted.

The analysis in this chapter is part of a larger body of research into how health claims became an object of government in the EU (Hendrickx, 2014). My understanding of the documents presented here has been corroborated through interviews with the authors, interviews at ILSI Europe headquarters in Brussels and participation in nutrition science colloquia. Organized as conversations, these interviews have allowed me to check the correct understanding of technical terms and to trace the history and rationale of these now widely cited publications. My aim, however, is not to simply describe this rationale, thereby running the risk of reiterating ILSI's vision. The challenge here is to conceptualize ILSI's strategy using my own means as an anthropologist and STS scholar, with the aim of taking a certain position, just like ILSI does – rather than a view coming from nowhere. Conceptually, I draw on literature in STS and the history of science. Of particular relevance to my analysis is the work of the two Belgian philosophers, Isabelle Stengers and Vinciane Despret.[1]

To begin, I briefly discuss ILSI's self-representation and how the organization is structured. I then move on to one of ILSI's most important and oft-cited publications on the subject of functional foods and health claims: the FUFOSE Consensus Document. I outline the nature of this publication and its strategic role for ILSI along with their conception of nutrition science. The main part of the chapter will consist of a discussion of ILSI's main nutritional concepts presented in the FUFOSE Consensus Document. The aim is to characterize these concepts as either scientific or something else. To complete the picture of ILSI's strategy, a follow-up project will be briefly discussed in the next paragraph, extending the operational scheme established by the FUFOSE Consensus

Document into the realm of policymaking. In the final section, I wrap up the main conclusions of my analysis in relation to the concepts of imagination and imaginaries. What are the *consequences* of the imaginary work that I have described and what does this entail in practice for social analysis?

ILSI Europe

I have been rather sparse with information as to the how and why of health claims in the EU: what are they, why are they defended, when did they emerge and why. In short, I have not placed health claims in their context so far. It is customary to draw up the context of a question to readers before delving into the specificities or technicalities of it. Such an approach is conventional, for reasons of clarity and pedagogy. At the same time, such an approach runs the risk of reiterating what others have defined as the 'context' for health claims. Indeed, drawing up a context for health claims and defining what they are form part of the technology of demonstration, designed to make health claims on food products possible. Health claims have emerged together with their definition and a rationale for their pertinence and necessity. Rather than separating context and subject matter, I propose to discover the two together through a scientific paper that stands as a landmark for the concept of functional foods and health claims *in the EU*. It is important to emphasize 'in the EU', as this is another instance, as we will see, of how subject matter and context inform each other. Put differently, I will analyse a context-building document, or the *setting of a problem*. A concrete entry point into that document to start with is its authorship: a consortium of life scientists interested in nutrition. Their texts are geared to turn their interest in nutrition into a specific nutrition science that they call 'functional food science'. The consortium was gathered by the European branch of ILSI. This is how ILSI presents itself on its website:[2]

> Founded in 1978, the International Life Sciences Institute (ILSI) is a non-profit, worldwide foundation that seeks to improve the well-being of the general public through the advancement of science. Its goal is to further the understanding of scientific issues relating to nutrition, food safety, toxicology, risk assessment, and the environment. ILSI is recognised around the world for the quality of the research it supports, the global conferences and workshops it sponsors, the educational projects it initiates, and the publications it produces. ILSI is headquartered in Washington, DC. It is affiliated with the World Health Organization (WHO) as a non-governmental organisation and has special consultative status with the Food and Agriculture Organization (FAO) of the United Nations.
>
> (ILSI website)

ILSI's 1978 founding father was Alex Malaspina, then vice-director of the Coca-Cola Company in the United States. He is said to be a 'real visionary'.[3] He created ILSI because Coca-Cola had a caffeine problem: rumours abounded that caffeine had toxic effects (see e.g. Troyer and Markle, 1984). Malaspina was determined to counter such rumours with scientific evidence. It *had to be* shown, with scientific arguments, that caffeine was not toxic. Eight years later, in 1986, ILSI created its European branch, 'ILSI Europe', in Brussels. In total, the organization counts 15 regional branches all over the world.[4]

ILSI is entirely funded by the food, chemical and biotech companies that pay for ILSI membership. Only the larger companies can afford to pay the membership fees, and therefore very few small or medium-sized enterprises are part of the ILSI network.[5] ILSI Europe gathers scientists from all over Europe into thematic task forces. Currently, there are 21 task forces,[6] covering themes such as: 'dietary carbohydrates', 'threshold of toxicological concern', 'addition of nutrients', 'consumer science', 'probiotics', 'prebiotics', 'novel foods and nanotechnology', 'packaging materials' and 'functional foods'.

The leading branch working on functional foods within ILSI's global network is the European one. As paying members, food and chemical companies such as Danone, Nestlé, Unilever, Kellogg's, Cargill, BASF and many others have the initiative of proposing a new task force.[7] The agenda of the task forces – the topics and problems that they address – is also set by the paying members. The task force is comprised of representatives of member companies and 'non-industry experts whenever required'.[8]

Functional food science for Europe

ILSI Europe successfully applied for a European grant from the European Commission's 4th Framework Programme, and in 1995 it started to extend its network through a project called *Functional Food Science for Europe* (hereafter 'FUFOSE').[9] This project resulted in the publication of a programmatic paper for functional food science in the British Journal of Nutrition in 1999. The paper has now become a landmark reference to cite in any report, paper or article that deals with functional foods, its science or its history.

The aim of the FUFOSE project was to 'establish a science-based approach for concepts in functional food science'.[10] In fact, as I show below, what is established is functional food science *itself*. This was to be done by setting up an international and multidisciplinary European network of researchers in the life sciences. More specifically, the objectives of the project were:

1 to assess critically the science base required to provide evidence that specific nutrients and food components positively affect target functions in the body;
2 to examine the available science from a function-driven perspective rather than a product-driven one; and
3 to reach consensus on targeted modifications of food and food constituents, and options for their application.

(Diplock *et al.*, 1999, p. 1)

The ILSI think tank starts thinking from the answers and solutions they already have. The *answer* is that 'specific' nutrients and food components positively affect the human body.

The aim, then, is to provide the right question or problem to the answer they already have: a science is needed to support the answer that specific foods are good for you. Science is to be subjected to the requirements of the provision of evidence. This requirement is confirmed in the third point: a consensus must be reached to align all researchers in the project on the problem that will accommodate the proposed solutions. The second point of the quote above deserves special attention: a 'function-driven perspective' is preferred over a 'product-driven' approach. By this, ILSI means that *specific foods and their nutritional composition* will *not* be investigated, but only the function of isolated nutrients in relation to functions of the body (hence 'functional' food). To understand this, consider that certain products, like yoghurts, are easier to promote as beneficial than snacks that are rich in sugars and fat, or soda drinks: yoghurts have a more favourable 'nutritional profile' and are potentially more credible bearers of health claims. Yet, ILSI needs to find an inclusive approach for its 62 different member companies. A second and related reason for eschewing a 'product-driven' approach is Europe's market itself. The composition of food products is a problematic issue that often divides the member states, one that risks obstructing the free movement of foodstuffs within the Community. It is important to note that what ILSI calls 'science', a 'science base' or a 'functional food science' is a strategic move to encourage and anticipate Community legislation for health claims as a promotional tool. In Europe and worldwide, health claims became very much in vogue during the 1990s. However, there were no common rules to deal with them. Each member state had its own policy.[11] Aware that certain countries were sceptical about this health niche, ILSI wanted to benefit from a first-mover position by proposing a science to evaluate health claims in terms that were favourable for the food industry at large. The core of the strategy was to *detach* 'functional components' from actual products and their complex and ambivalent nutritional composition, and to *detach* 'target body functions' from an individual's complex clinical experience. This detachment is also a general operation to transform products with particular histories into goods that can circulate in the calculable space of markets before attaching themselves again to particular people who buy them (Callon, Méadel and Rabeharisoa, 2002). The detachment that ILSI proposes, however, is peculiar: to value and qualify food *ingredients*, food *products* must become invisible. Historian and philosopher of science Gyorgy Scrinis has coined such a strategic focus on separate nutrients 'nutritionism'.[12]

A final aspect that I would like to highlight before looking into the contents of functional food science and the Consensus Document is ILSI's *claim* – in the sense of claiming a right – to speak in the name of public health. ILSI wants to contribute to the improvement of public health through the advancement of science. I have already indicated that ILSI works on several topics, ranging from food benefits and safety to environmental safety and sustainability. Here I focus on functional food science:

> We are at a new frontier in nutrition science because, at least in the industrialized world, concepts in nutrition are changing significantly. We are progressing from a concept of 'adequate nutrition' to one of 'optimal nutrition'. We have moved from a former emphasis on survival, through one of hunger satisfaction and of food safety, to our present emphasis on the potential for foods to promote health, in terms of both improving well-being (mental and physical conditioning) and reducing the risk of diseases.
>
> (Consensus Document, p. 5)

What this excerpt shows is that science is expected to be able to help 'us', in the industrialized world, to eat better. ILSI, speaking in the name of public health, defends nutritionism as a timely solution, as Western societies face the problems of ageing, chronic and non-communicable diseases like diabetes, obesity, cancer and cardiovascular diseases (e.g. Roberfroid, 2000; Howlett, 2008; Chadwick *et al.*, 2010).

In its mission and faith to connect science and public health, ILSI and its nutrition community envisages a form of specialism or even professionalism through a common approach to health around the object of functional foods (instead of drugs): 'Functional food science is a new discipline that is part of the science of nutrition and is aimed at stimulating research and development of these foods by using a function-driven approach,' writes the scientific coordinator of the FUFOSE project (Roberfroid, 2000, p. 1661). However, in contrast to past communities of advocates for more rigorous drug trials in the form of the randomized clinical trial (Marks, 1999), the community that the FUFOSE project created is not anti-commercial. The boundary that this community wants to set is not between private and public but *sectorial*: between the agri-food sector and the pharmaceutical sector and their respective markets. 'In no case must functional foods be considered drugs,' write the authors of an article in the *Scandinavian Journal of Nutrition* (Coppens *et al.*, 2001). The authors, at that time, were all members of the Health Claims Expert Group of the European Confederation of Food and Drink Industries (CIAA).[13] If Marks's therapeutic reformers operated according to a certain moral economy with the authority of evidence as its point of gravity (Marks, 2009), then, I argue, the functional foods community is trying to establish a point of gravity and a form of evidence of its own, in a moral economy where 'stakeholders' sit together peacefully, agree on the problems of a given, common world and forge consensus on the solutions under the sign of science.

Qualifying nutrients and body functions

The FUFOSE project ended in 1998, and in 1999 the results of their final consensus meeting were published in the *British Journal of Nutrition* (Diplock *et al.*, 1999), along with the following working definition for functional foods:

> No universally accepted definition of functional foods exists. In fact, because functional foods are more of concept than a well-defined group of food products, a working definition is preferred for the purposes of this Consensus Document.
>
> A food can be regarded as 'functional' if it is satisfactorily demonstrated to affect beneficially one or more target functions in the body, beyond adequate nutritional effects, in a way that is relevant to either an improved state of health and well-being, and/or reduction of risk of disease. Functional foods must remain foods and they must demonstrate their effects in amounts that can normally be expected in the diet: they are not pills or capsules, but part of a normal food pattern.
>
> (Diplock *et al.*, 1999, p. 6)

Let us consider the definition in separate parts. Here is the first part:

> A food can be regarded as 'functional' if it is satisfactorily demonstrated to affect beneficially one or more target functions in the body, beyond adequate nutritional effects, in a way that is relevant to either an improved state of health and well-being, and/or reduction of risk of disease.
>
> (Diplock *et al.*, 1999, p. 6)

The starting base is 'food' but the most important word in the citation is 'demonstrated'. Functional food is a concept for *food that has passed a test* of some sort. The kind of test remains unspecified in this definition. Functional food is not an object or a predefined group of food products, as the authors argue, but is rather a *qualification* of food ('a food can be regarded as "functional" if ...')

The authors then repeat that only foods can be qualified as functional:

> Functional foods must remain foods and they must demonstrate their effects in amounts that can normally be expected in the diet: they are not pills or capsules, but part of a normal food pattern.
>
> (Diplock *et al.*, 1999, p. 6)

Pills or capsules cannot participate in the kind of test that attributes effects 'beyond nutrition' to foods. The question then becomes: what are food's effects beyond nutrition, and how do you demonstrate them? The main issue in the Consensus Document is to make a number of proposals to answer that double question. The Document in fact proposes the term functional food and its definition as a working programme. To better characterize this programme, let me reformulate the definition of functional food in terms of the *genus–differentia* definition it implies.[14]

The definition specifies a genus (functional food belongs to the category of food), and a differentia (it affects functions in the body beyond nutritional effects). The differentia is hypothetical in nature and refers to other concepts that need to be defined ('target functions in the body', 'beyond adequate nutritional effects', 'improved state of health'). Now, one could argue that this differentia does not really distinguish functional food from other food: all foods are likely to have effects beyond mere nutrition, it is just that we are not aware of it. But this is exactly the point: the Consensus Document is not about a new category of food, but about what I would call a methodology of *demonstration*. Both the genus and differentia, if we follow the logic of the Consensus Document definition, refer to the effects of food and not the essence of it. The aim of the Document is to propose a method, or rather a series of methods, to demonstrate these effects. What complicates things is that these methods are proposed together with new concepts: effects, qualified as 'beyond basic nutrition' have to be made visible in 'target body functions', indicating 'an improved state of health'. So the Document's working definition does not qualify as a referential definition since the description does not refer to known terms. Rather, it is an operational definition or a working programme for *functional food science*, which I describe as *a science that is capable of demonstrating what food is capable of in human bodies*. The structure of the article gives a hint of what this programme looks like. Here are titles of the most important subsections in the introductory part of the article: 'From traditional to new concepts in nutrition. From new concepts in nutrition to functional foods. From functional foods to functional food science' (pp. 5–6).

If it seems a bit awkward to have a science *following from* the functional foods concept, and not the other way round, it must be remembered that the 'science' described in the Document is in fact a method of demonstration, subject to specific constraints. 'Functional food', as a concept, is defined, I argue, *according to the requirements of the type of human intervention trials that are likely to make effects of certain nutrients in the human body visible*.

The Consensus Document introduces a definition of functional foods that is dependent on methodologies yet to be developed. Only these methodologies can call functional foods into *existence*.

Figure 4.1 helps us to understand what the methodologies that define and qualify functional foods consist of. A functional food component is anything that can be formulated into a food matrix and that is safe for consumption. The food component can be a nutrient, like a vitamin or a mineral, but also a non-nutritive component like a bacterial strain, certain fibres, or plant sterols (e.g. in cholesterol-lowering margarine). The authors propose themselves to investigate the 'black box' between a food component and a health-related outcome. The idea is that there is a chain of markers that mechanistically lead to an outcome. The original explanation for Figure 4.1 is:

> This is a diagrammatic representation to show how different types of markers would be expected to lie within a logical progression from the food component to the health outcome. The types of markers are completely independent of each other. Markers can either be indicators or, if they can be proven to be causal, factors.
>
> (p. 7)

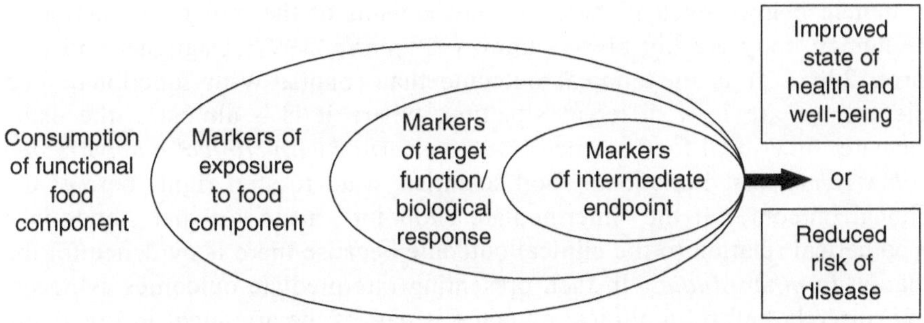

Figure 4.1 Classification of markers relevant to the effects of functional foods.
Source: reproduced from Diplock *et al.*, 1999, p. 7.

The first type of marker is the marker of *exposure*, for example a serum, faecal, urinary or tissue marker. These are 'available' and 'feasible' markers, as the Document indicates. The increased level of red blood cell folate is a marker of the exposure to folate in food, for example. The second type of marker is a marker of *target function* or biological response. The example given is the reduction of plasma homocysteine as a possible response to dietary folate. The third type of marker is the marker of *intermediate endpoint*. This concerns, again, the measurement of a biological process that is somehow related to a clinical outcome (improved state of health and/or reduced risk of disease). The example given is the extent of narrowing of the carotid artery as evidence for cardiovascular disease. Another example would be the lowering of cholesterol which is also considered to be related with cardiovascular disease.

Two key issues about this classification of markers are of importance for my characterization of functional food science. First, the word 'clinical' has no significant role in this Document. The word occurs four times over the 27-page article, but in three of the four cases it has no significant relation to the methodologies of functional food science. Only one of these four mentions is pertinent to this study. In a paragraph on 'safety considerations' the following is stated:

> Protocols for human nutrition studies need to be developed including, in some cases, post-marketing surveillance. Even though the design of *clinical studies as used in drug development* can serve as a reference point, specific protocols and specific criteria relevant to functional foods might be needed.
> (p. 8)

I have emphasized in which context the word 'clinical' occurs: in relation to drugs. Drug testing, both with regards to safety and efficacy, involves the evaluation of clinical outcomes or 'endpoints'. A clinical outcome is a therapeutic or toxic effect that is clinically observable: a human being recovers, gets better, heals or, on the contrary, gets sick or worse. For the development and putting on the market of drugs, it is crucial to know the effect they have on human beings, even if the exact mechanisms of the molecule's action in the human body are not always known (Pignarre, 1997; Dagognet and Pignarre, 2005). It is the clinical outcome that counts. With functional food science, we are in a different situation, where it is – almost – the exact opposite: functional food science *is not concerned with clinical outcomes, but with mechanisms*. Functional food scientists want to stop right 'before' the clinical outcome, at the 'intermediate endpoint', a marker that stands in a hypothetical relation to the clinical outcome because there is evidence for the relation in *other studies*. In fact, presenting intermediate outcomes as occurring 'directly before' a clinical outcome is part of the argument in functional food science: it suggests a straight *causal* pathway of bodily responses that has predictive value. One of the main authors of the Consensus Document has developed this idea further and calls it 'evidence-based mechanistic reasoning' (Aggett, 2011). The reason to encourage this type of 'reasoning' and to focus on markers is to avoid any confusion with drugs and the costly business of large-scale clinical trials.

This brings me to the second point that characterizes functional food science as a specific practice. When I said that functional food science is concerned with mechanisms (rather than clinical outcomes), the understanding of mechanisms is not a goal in itself. Functional food science is a science that demands *leeway* and looks for associations rather than strong causality. If causality can be established, then all the better. If it cannot be established, then this is no reason to say that a claim about nutrient's health benefits cannot be formulated:

> The differential classification (of markers) is considered to be of real importance in the development of new markers for use in human studies. The result from such studies can also form a scientific basis for formulating and controlling claims.
>
> (Diplock, 1999, p. 8)

The term 'indicator' is used for markers that are not causally related to the health outcome, and 'factor' is used for those that are causally related, although it is not specified what causality exactly means.[15]

Criteria for markers are proposed in bullet points such as:

- Markers should represent relatively immediate outcomes, which can be used to assess interventions in a reasonable timescale; they could, therefore, wherever possible, replace later and more remote outcomes as have been used in some epidemiological studies.
- Markers must be rigorously validated and amenable to standard quality-control procedures.
- Markers must be clearly linked to the phenomena involved in the biological process being studied....

...

- Markers must be measurable in easily accessible material, or obtainable using methodology that must be both ethical and minimally invasive.

(Diplock, 1999, p. 8, my emphasis)

These criteria constrain the possibilities of scientific research, while insisting on the fact that it must be rigorous and quality assured. Functional food scientists, then, appear as a type of *technician* constrained to avoid too difficult (and time-consuming) questions and focus on *solutions*.

A final remark about the concept of 'markers' in the Consensus Document, and their classification in 'indicators' and 'factors'. This terminology is used in the Consensus Document to avoid the term 'biomarker'. As a later publication (Biesalski *et al.*, 2011) of ILSI Europe-affiliates explains,

> The FUFOSE consensus eschewed the term biomarker, which has biochemical connotations. It preferred markers because this acknowledged the broad range of available and applicable markers. Thus, apart from biochemical markers, study outcomes can quite feasibly be derived from, among others, behavioral or psychometric outcomes, physiologic performance, adaptive phenomena, and metabolic clearance studies;... In fact, any marker that can be quality assured and validated should be appropriate for establishing causality. Often, it is likely that a 'battery' of markers might be needed to address the perspectives of evidence appraisal ... for multiple and variable sources of data. In fact, given the quality of data available for health claims, these considerations about markers should inform 'new human intervention studies using appropriate markers to generate readily interpretable, valid and reliable data'.
>
> (Biesalski *et al.*, 2011, p. 4)

To summarize, the category of 'markers' must be broader than the biochemical 'biomarkers'. This diversity of markers must also be combinable (it is not clear how) to generate 'batteries' of markers. These terminological and conceptual choices are all related to the fact that functional food science is not a fundamental science, but a strategic programme to generate evidence for health claims. FUFOSE was written and published before a European regulation on health claims existed, at a time when food companies and national authorities were pushing for a regulation of claims on food products. Food companies were pushing to have access to a European market with a clear and light approval procedure on the European level. National authorities were pushing because some of them were put under pressure by industry.[16] When the Consensus Document was written, there were at least three reasons to propose 'markers', subdivided into indicators (correlation) and factors (causality). The first is to *avoid being limited to* biochemistry; the second (related to the first) is to *broaden the scope* of marker-oriented research to cognition, psychology and performance; the third is to *broaden the evidence-base* for functional food by including weak causality or correlation.

Let me sum up the characteristics of functional food science, as a specific practice, that we have explored up to now. I will call the first characteristic 'the externalization of the clinical': the clinical is suggested but at the same time kept at bay. Instead of 'clinical studies', the terms 'human studies' or 'human intervention studies' are used. The clinical is situated outside the box of markers and mechanisms that functional food scientists would ideally like to investigate. *Investigating* is perhaps not the right word. *Operationalizing* is probably better, in view of the definition that I have proposed for 'functional foods' or 'functional food science': a technology of demonstration, established *according to the requirements of the type of human intervention studies that are likely to make effects of certain nutrients in the human body visible*. The human intervention studies, as we have seen, are in fact marker-oriented studies, with leeway for associations or 'batteries' of markers to approximate causality. The idea is defended that association *as such* has a cumulative effect and increases evidence for a relationship with the clinical. This is the second characteristic of functional food science: it excludes the risk of association or 'articulation', which is *constitutive* in theoretical-experimental sciences (Latour, 2007). The risk taken in scientific practice is that the object under study is *given the power to redefine the questions addressed to it* (Stengers, 2006). To be sure, my argument does not concern the competence of nutritionists or life scientists here. Competence is not the issue. What I am concerned with here is what Stengers (2006) calls the *milieu* or environment that FUFOSE and ILSI create for scientific research. This environment doesn't allow researchers to pose their own questions and take the risk of granting objects the power to redefine those questions. The Consensus Document formulates a series of concepts and constraints that do away with this risk, because the final aim is not to gain fundamental knowledge on a food ingredient or body function but to *pass through the human body as a site of demonstration of the ingredient's market value*. If this resembles the way drug molecules are socialized and put on the market (Pignarre, 1997), the functional food science imaginary turns the body into a site to differentiate food from drugs, and health from disease, by externalizing the clinical from its 'box' of mechanisms. Functional foods have no patients, but consumers (cf. Mol, 2008). Consumers' experience of well-being is irrelevant to the market test that functional food science proposes. With regards to 'science', the question that is asked in the Document is: 'how much science is enough?' (Heasman and Mellentin, 2001). Figure 4.2 below is also reproduced from the Document and I propose to read it as follows: this figure visualizes the body, detached from clinical experience, as a point of articulation between a food component and a *claim* with market value.

Again, the difference with scientific practice as a mode of taking risks with objects is salient. Making a scientific claim means taking a risk: the claim will be examined by colleagues and the object itself can also at a certain point contradict a claim or hypothesis. In *Laboratory Life*, Bruno Latour and Steve Woolgar (1986) show that claims or statements go through a spectrum of modalities, and that these modalities are also a resource for other scientists to make the statements of their colleagues more or less credible. Some claims will eventually

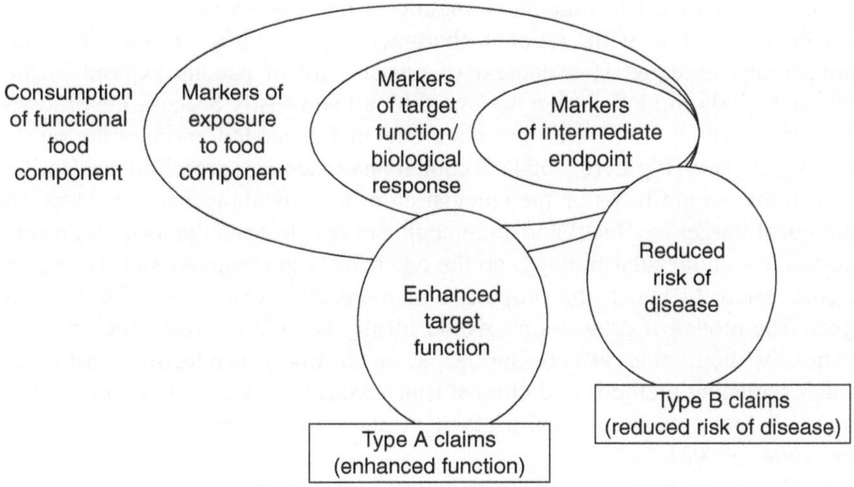

Figure 4.2 The human body as a point of articulation between components and claims.
Source: reproduced from Diplock *et al.*, 1999, p. 25).

'descend the ladder' and be laid to rest as claims. Other statements, however, will become more and more solid. A scientific object is a solidified claim: science creates its own objects, which represents a considerable investment and the taking of risks. Functional food (science) uses the risks taken by other researchers; it starts from recognized objects and makes them subordinate to a claim. The object comes to *carry* a claim, and this claim is the final result and benchmark of excellence. The logic with respect to *Laboratory Life* is turned on its head: instead of making risky claims to articulate (or not) objects, the question becomes what kind of fixed procedure can deliver good claims without them mattering for the object. In that sense, and to avoid any further confusion with science, I propose here to call functional food scientists *claim technicians*. This is not pejorative per se, but it enables us to distinguish the 'technical' from the scientific in the way that Isabelle Stengers proposed: '[T]o treat technique as distinct is crucial if we are to resist the effects of fascination associated with discourse about the inexorable technoscientific redefinition of the world.' (Stengers, 2011, p. 340). The scientific and the technical do not automatically imply each other, and distinguishing them draws attention to the specific conditions their interaction requires.

As I have discussed earlier on, the Consensus Document announces a new frontier in nutrition science that would help us to eat better. This is the 'inexorable technoscientific redefinition of the world' in the quote, the fascination of which must be resisted, because it is a new frontier not in science but in food technology, which is confirmed by some of the authors of the Consensus Document themselves.[17]

Functional food science is designed to craft a territory for therapeutic claims outside the scientific and regulatory regime of pharmaceuticals, but it actually has a lot in common with 'rational pharmacology' or 'drug design'. Rational pharmacology emerged in a context of crisis: a lot of patents expired in the 1980s and 1990s and innovation became difficult and costly because the production of drugs had, until then, followed a model that generated 'more of the same' over several years (Pignarre, 2004). Rational pharmacology promised a scientific approach that would focus on the elucidation of biological mechanisms between a biological 'target' or 'lock' and a molecule or 'key' to open the lock. A lot was expected from molecular biology, on the one hand, and computer simulation, on the other hand, to match the properties of molecules with those of biological targets. The model of drug design would ideally be able to create such perfect matches, without side effects and at a much lower production cost than through screening methods and clinical trials. Moreover, rational pharmacology imagines a future where clinical trials themselves are no longer necessary (Pignarre, 2004, p. 103).

The Consensus Document presents a nutritionists' variation on the project of rational pharmacology or drug design. It is not called 'rational nutritionism' but 'evidence-based nutrition' or 'evidence-based mechanistic reasoning'. Through the notion of 'evidence-based', however, it acknowledges the missing links in the chain of markers and therefore proposes, in the same run, a scheme to assess evidence. In other words, it proposes how regulators should deal with the incompleteness of claim substantiation: different kinds of studies, indications and markers should all be recognized as contributing to the strength of a health claims dossier and *should be taken into account*. Scientific proof should not be dependent on one specific demonstration, and certainly not on the strong causality of clinical trials. A certain amount of science should be enough, and that amount, preferably, shouldn't involve anything clinical, *especially* not the randomized controlled (or clinical) trial (RCT). By eschewing the term 'biomarker', functional food science, like rational pharmacology, places its hopes on molecular biology as an approach that will lead to the understanding of biological mechanisms, which is something that chemistry is less equipped for.[18] Next to molecular biology, the Consensus Document mentions immunology as a promising field for their purposes, as it offers dynamic concepts of immunity that are interesting for functional food science. It opens the door to new markers, not indicating a *state* of the organism but its ability to *adapt*.

Functional food science, then, as 'evidence-based nutrition', is a project similar to that of drug design, and it comes with its own policy:

The scheme below again indicates the special nature of functional food technology as a form of regulatory science (Irwin *et al.*, 1997). In fact, the need for a 'scientific basis' is stipulated as a starting point. Such an assertion is absurd within scientific practice, but it makes sense within product development. Type A claims are *enhanced function claims* and Type B claims are *reduction of disease risk* claims. The European Commission's proposal for health claims regulation in 2003 (four years after the publication of the Consensus Document

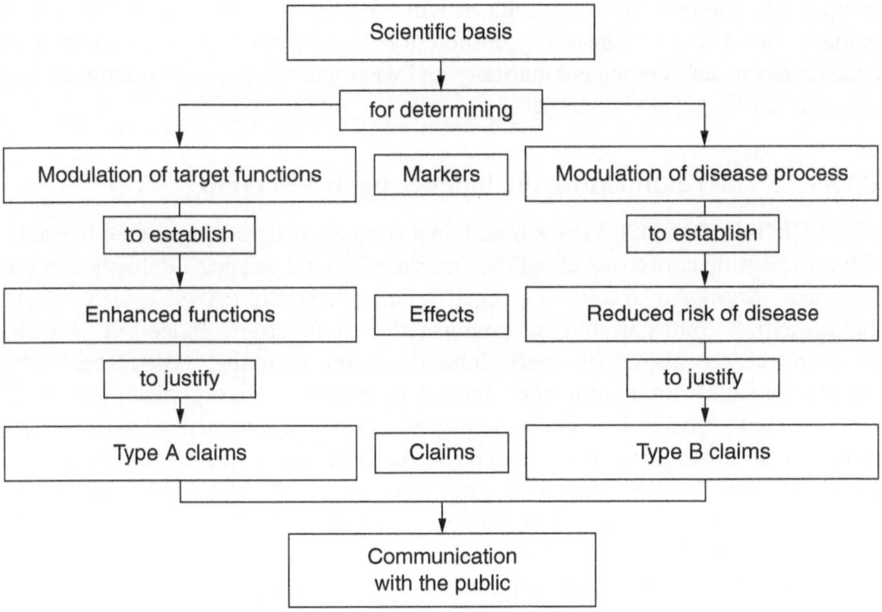

Figure 4.3 A functional food road map to product development and regulatory science.
Source: reproduced from Diplock *et al.*, 1999, p. 25.

in 1999) took up these categories and cited FUFOSE as 'valuable work' (European Commission, 2003, p. 9). In that same proposal, the Commission interpreted the legal prohibition on medicinal (or 'disease-related') claims on food as going too far, and as not in keeping with new technological developments (2003, p. 2). This made disease *risk reduction* claims possible. That way, they avoid connecting a food ingredient to disease as a clinical outcome, while still making a therapeutic claim. Avoiding that confusion also fits the food industry's agenda to create a market niche out of 'health'.

FUFOSE's follow-up project, conveniently called PASSCLAIM – also funded by the European Commission and coordinated by ILSI Europe (2001–2005) – proposed criteria to assess health claims made on food products, thereby providing a source of inspiration for the European Regulation on Nutrition and Health Claims (NHCR), which was under discussion when the PASSCLAIM results were published in the *European Journal of Nutrition* in 2005 (Aggett *et al.*, 2005). The project ran from 2001 to 2005 and was already mentioned in 2003 for its 'considerable work' in the Commission's first proposal for a NHCR (European Commission, 2003); it is still recognized as an important work for policymakers today.[19] PASSCLAIM is an acronym for the process for the assessment of scientific support for claims on foods and it is one step closer to policymaking – or, rather, the premarket approval of functional food – than FUFOSE. FUFOSE, as we have seen, introduced the concept of functional foods with a specific terminology (functional components, target functions, health benefit, marker, etc.) and a classification of markers. But it didn't address the question of how different types of studies should be *assessed* by any competent authority and what their respective importance is. I will now briefly turn to this project.

PASSCLAIM: calibrating the balance before weighing

Like FUFOSE, PASSCLAIM is based on a strategic programme geared to establishing scientific consensus about the criteria of 'sound science'. Published in the *European Journal of Nutrition* (Aggett *et al.*, 2005), the PASSCLAIM project gathers expert groups around pre-established body functions. Indeed, ILSI itself, as we have seen above, has been subdivided into thematic 'task forces'. The themes are set by food companies that are members of the organization.[20] Even more explicitly than in FUFOSE, the PASSCLAIM project, as the acronym indicates, is a strategic programme to deliver a science base to policymakers when a European regulation is under way. More precisely, PASSCLAIM wants to set the terms of such a science base. The Commission, which was not sufficiently staffed to deal with the increasing number of technical issues that the EU decided to regulate,[21] welcomed such input, and the framework programmes were there to generate policy-relevant inputs.[22] A total of six criteria for the scientific substantiation of claims were proposed, with a number of subdivisions. I will restrict myself to a number of issues concerning 'human data', to be generated from 'intervention studies'.

First of all, in contrast to FUFOSE, more attention is given here to the importance and strength of the randomized clinical trial (RCT). However, the main argument is consistent with the line of thinking that FUFOSE initiated: *all* available data should be taken into consideration, without granting absolute privilege to one type of study (e.g. RCTs). In the words of the authors, 'For all studies and methodologies, quality and power may take precedence over the type of study in weighing evidence for the substantiation process' (European Commission, 2003, p. 14). This statement, resembling a principle, is complementary to the content of criterion six, which says that 'A claim should be scientifically substantiated by taking into account the totality of the available data and by weighing the evidence'. A similar reflection is made with respect to the elucidation of biological mechanisms, which is the ambitious ideal of functional food science, but not for health claims: 'For the substantiation of a claim, it is ... more important to demonstrate a consistent effect of a food or food component on health across a range of studies than to have a scientifically substantiated mechanism' (ibid., p. 15). In other words, health claims *cannot wait* until science advances. In fact, the elucidation of mechanisms on the one hand and RCTs on the other seem to represent two extremes in terms of requirements, none of which functional food scientists wish to privilege when it comes to substantiating health claims. PASSCLAIM encourages the mixture, association and accumulation of different studies, depending on what is available. I already pointed to this in FUFOSE: the *risk* that scientists usually take when associating and articulating their experiments with the objects of investigation is reduced in advance as much as possible. Such scientific and financial risk is at a minimum when computer models can be used: 'Many laboratory or computer-based models are now used in nutrition to circumvent the long and costly procedure of human studies, to define mechanisms and predict behaviour in biological systems. Such models can provide additional evidence for the substantiation process' (ibid., p. 14).

For nutritionists, computer simulations are indeed of some interest, as they address behaviour in biological systems, rather than behaviour in *humans*. Functional food scientists know that conducting RCTs, the gold standard for the testing of drugs (Timmermans and Berg, 2003), are not always easy to design for food components. The authors discuss these difficulties, while at the same time giving some recommendations on how to conduct RCTs, or other human studies, as well as possible.

A first concern is the selection of 'subjects' to participate in a study:

> Subjects should be selected on the basis that the appropriate control group is one with a typical diet [*sic*], and not a special diet that might interfere with the intended benefit. For example, it might not be appropriate to use vegetarians to test the effect of an added fibre.
>
> (European Commission, 2003, p. 16)

More generally, the study group's background diet and lifestyle must be characterized to exclude confounding factors. Put differently, beneficial nutrients, in order to be nutrients and not drugs, must be demonstrated to have an effect in dosages that can be expected in a human diet. At the same time, nearly all confounding variables stem from the human diet. Humans, as the authors concede, are exposed to many active substances in their diet. So, which of the tested active substances is *already present* in the diet and how can we know about it? And what about interactions of nutrients in the diet with the tested substance? Diet must then be normalized through the selection of subjects (and the exclusion of vegetarians), and by the invention of devices to ensure the subject's compliance, especially where markers of intake or exposure do not exist. A questionnaire or logbook can be used as a device to check what people eat and if there are changes in the 'baseline diet', characterized at the beginning of the study. However, the authors go on, *misreporting* is a well-known problem, especially in 'obese subjects', who tend to underreport what they eat (ibid., p. 17). In addition, the act of recording is thought to influence food choices and intake (ibid., p. 17).

These examples show that 'taming' food through clinical trials, in order to obtain clear responses from it, implies that human beings must be tamed to a large degree as well, without any certainty as to the final results. One could conclude that there is not much sense in doing such trials, given the many confounding factors of substances that are already present in the normal diet of many people. But ILSI draws another conclusion: if such trials are complicated, then the clinic must be detached from the body and markers made to stand for the clinic. Markers, they assert, can be handled. Markers can act as a reliable witness to exposure on the one hand, and as a site of inscription for predictive therapeutic scenarios on the other.

Re-engaging imagination

The meanders of FUFOSE and PASSCLAIM tried to tackle what is at once an economic, political and ontological problem: the human body as a *reliable witness* (Stengers, 2006) for both the truth-claims and the marketability of a product. The fact that the human body is also a testifier for drugs is relevant. If functional food must pass through the human body before being put on the market, *just like drugs*, then the human body becomes the site to create a difference with drugs. For ILSI's nutrition scientists, academic and industrial, functional foods must become agents *in their own right*: agents of *health*. The food industry wants to address health *on its own terms*, with its own agents, and these agents require a new nutrition science – functional food science – with its own methodologies, addressing objects worthy of specialized, autonomous expert judgement. Evidence-based nutrition, present in FUFOSE and PASSCLAIM and explicitly championed as opposed to evidence-based medicine in more recent papers (e.g. Biesalski *et al.*, 2011), is a conceptual apparatus to invent both a new type of product and an expert to evaluate it.

The ILSI consortium claimed that we are at a new frontier in nutrition science, and suggested that a new role for food in public health has become possible. However, the stratagem analysed is not oriented towards new research questions and possibilities in science, but towards ways of *closing down* scientific questions and speculations. ILSI wants to create not only certainty but a *guarantee* that their methods and concepts will bring certainty with respect to the relationship between food ingredients and the human body – a certainty that is translatable to a health claim on a food label. Scientific research arguably also aims to bring more certainty, but the crucial difference with ILSI's evidence-based and consensus-based approach is that scientists will never guarantee that a specific method will bring that certainty. Revising methods and concepts are part and parcel of scientific work and nothing can be guaranteed in advance. Arguably, then, the guarantee that ILSI wants to bring for its own corporate, commercial, political and sectorial reasons signals the end of scientific imagination. Does this mean that ILSI is unimaginative? The document I have analysed, full of strategic concepts, seems to point out the opposite. Where to go then, with this concept of 'imagination'?

The theme of this book is the relation between science and imagination or imaginaries in shaping the future. Analysing the ILSI stratagem, where imagination is being put to work into specific operational concepts, has convinced me that more precise conceptual distinctions in the realm of what we call 'imagination' and 'imaginaries' may be needed. This, in relation to a distinction that has been central in this chapter: the distinction between scientific practice on the one hand, and the development of policy-oriented technical concepts, or an *evidence-base*, on the other. This distinction is critical in order to avoid and resist the expectations, rhetoric and intimidations that result from the quick merging of science and technology. This merging obliterates the different relations and exigencies vis-à-vis an object or phenomenon under research, as I have tried to show through my close reading of functional food 'science'. The *shibboleth*, as Latour calls it (1997, 2004), that distinguishes science from – in our case – a technology of demonstration wrapped in the discourse of science, is the following question: 'is the object under research granted the power to redefine the questions that we address to it?' A good scientific experiment, then, 'maximizes the occasion for the phenomenon at hand to raise its own questions against the original intentions of the investigator' (Latour, 2004, p. 219). The approach, the exigencies and the consequences of scientific *practice*, then, are very different than the approach that ILSI pursues *in the name of* science. In this conclusion, I suggest that this difference should make us attentive to different forms of 'imagination' or collectively cultivated 'imaginaries'. ILSI, arguably, mobilizes a form of imagination to operationalize the human body as a test site different from the clinical trials for drugs. As we have seen, this working of the imagination operates under a set of very specific constraints that are at once legal, political, technical and commercial. As such, the imagining work relates closely to already existing arrangements that are actively remodelled. Put differently, ILSI *engages* with its object in a very specific way. Common dictionaries define imagination in contrast to reality: imagination is a form of fiction. However, what both scientific practice and evidence-based approaches like ILSI's have in common is that they are geared to *change* reality, thereby operating in the interstices of fiction and reality. What makes them different is the way they engage with the world in order to produce new realities.

If we take 'imagination' to cover very different *modes of engagement* with the world and with specific phenomena and objects, then we might also want to become attentive to the forms of imagination that we deploy as social scientists. And this stands in direct relation to the status of an analysis like the one in this chapter about the specific strategies of a group like ILSI. Concluding that ILSI is not a scientific consortium but a strategic think tank or lobbying group is hardly surprising.

So, rather than simply denouncing ILSI's claim to science while they are doing something else, I wanted to draw attention to the specificities of what it is they are doing. I wanted to tease out how they engage with their main object: functional foods. ILSI indeed provides the basis for what can be called 'regulatory science' (Irwin *et al.*, 1997) or 'cameral science' (Stengers, 2013), but the texts that I analysed also shed light on an ontological matter: the separation of food from drugs needs conceptual, material, methodological, political and technical *work*. Food ingredients, health claims and the human trial subject, but also drugs and pharmacology, are *boundary projects* (Haraway, 1997). Indeed, the division between food and drugs is a historical and ongoing boundary project. Considered that way, it is not helpful to see functional foods as a blurring of the division between food and drugs. Rather, functional foods are a concerted, consensus-based and evidence-based project to reinvent and reaffirm that division differently. Many commentators have characterized functional foods as an instance of the medicalization of food. Again, the opposite is true, as the main stake in the functional foods projects is to articulate a difference with medicine. This changes the game for the social researcher too: given distinctions are not suddenly blurring, and indeed distinctions are perhaps not 'given' at all. The difference between food and drugs, their markets and experts is not a given, but it is *at stake*. This establishes ontological uncertainty as the 'default position', as it were, from which to start thinking, posing questions and engaging different forms of imagination. This resonates with a line of inquiry that runs through the entire work of Michel Foucault. One of his main questions was what *singularizes* things (e.g. Burchell, Gordon and Miller, 1991). Here, we could ask, along with Foucault, what singularizes food and medicine, how are they singularized and how did we come to talk of 'the human body' the way we do? And even more importantly: what other versions of the human body exist, or *could* exist (i.e. could be imagined and realized)? Imagining boundary projects is inherently political, and the deal is not to denounce that political nature, but to resist the early taming and closing down of important societal matters by the hasty (self) appointment of experts.[23] Thinking from (and with) ontological uncertainty indeed denaturalizes divisions between the makers of knowledge, recipients of knowledge and a priori limitations on who is concerned by this knowledge. It emphasizes the singularization of problem-frames and the fact that things could have been different and could still be different. Next to critical genealogical analysis, it makes a bet on possible worlds, while being sensitive to the question of who is in charge of imagining those possible worlds. Thinking from ontological uncertainty relates knowledge to power indeed, while at the same time making an appeal to a kind of imagination that seeks to cultivate an appetite for such uncertainties and the speculative horizon of the possible.

Notes

1 In dialogue, both empirically inclined thinkers have elaborated a pragmatic and normative theory as to what the practice of science demands from its researchers, thereby redefining what a successful experiment means, other than in methodological and epistemological terms. Their work connects well to that of Bruno Latour, as can be read, for example, in an article that Latour has written on what he calls the 'Stengers–Despret falsification principle' (Latour, 2004). These relations and concepts will be developed and applied through the chapter.
2 www.ilsi.org/Europe/Pages/Who-We-Are.aspx [accessed 15 January 2014].
3 Personal interview with ILSI's deputy director in Brussels, 26 August 2011.
4 Argentina; Brazil; Europe; China (ILSI 'focal point'); India; Japan; Korea; Mexico; North Africa and Gulf Region; North America; North Andean; South Africa; South Andean; Southwest Asia Region; Taiwan. For more info and a hyperlink to each branch, see www.ilsi.org/Pages/GlobalNetwork.aspx.
5 Interview at ILSI, Brussels, 26 November 2010, with FM and FR.
6 As counted on the ILSI Europe website on 15 January 2014: www.ilsi.org/Europe/Pages/Task-Force-List.aspx.

7 The procedure is briefly explained on the website.
8 Ibid.
9 The full name of the project is European Concerted Action on Functional Food Science in Europe. FUFOSE is the abbreviation used in publications of the Commission and of the project members.
10 Consensus Document, preface.
11 See, for example, Heasman and Mellentin (2001).
12 See his 2013 book. For his characterization of nutritionism as an ideology, see Scrinis (2008). Michael Pollan (2008) has introduced the term 'nutritionism' to a wider audience in his best-selling book *In Defense of Food. An Eater's Manifesto*.
13 This is mentioned in the form of a disclaimer underneath the article. The disclaimer further states that: 'This article expresses the views of the authors and does not necessarily represent the policies of their employers.'
14 The genus–differentia definition goes back to Aristotle, and a discussion of it can be found in Granger (1984). Granger argues that Aristotle defines the relation between genus and differentia in three different manners, and that this represents three stages in Aristotle's thought. My use of the two terms is less nuanced and serves only as a first clarification of the term 'functional food'.
15 I refer again to cholesterol as one of many risk factors. Causality is mentioned three times in the entire document, without it being the main issue in the section where it occurs.
16 Interview at the European Commission, DG Sanco, Brussels, 23 April 2013.
17 Personal interview at ILSI's Annual Symposium, Brussels, 24 March 2011.
18 See Pignarre, 2004, p. 96, for the inversion of the traditional roles of chemists and biologists in rational pharmacology.
19 Interviews at the Belgian Federal Ministry of Health, Brussels, 12 March 2012, and at the Commission's DG Sanco, Brussels, 23 April 2013.
20 Interviews at ILSI Europe, Brussels, 16 November 2010, 26 November 2010, 26 August 2011.
21 Barry (2001). During the 1980s, the EU went through a reform and expanded its areas of regulation. See e.g. Young (2005, p. 105) and the work of lobbyists Guéguen (2007) and Guéguen and Rosberg (2004).
22 In the preface to the PASSCLAIM published article, European Commissioner Jürgen Lucas states: 'The results of PASSCLAIM have been and will be an important input into the regulation on nutrition and health claims made on foods, which is under discussion at the moment' Aggett *et al.* (2005, p. 3).
23 This, of course, echoes Bruno Latour's *Politics of Nature* (1999), and Callon, Lascoumes and Barthe's *Acting in an Uncertain World* (2009). I connect their work to Haraway's commitment, through all of her work, to speculation, fabulation and science fiction.

References

Aggett, P., 2011. From PASSCLAIM to health claims. In: ILSI Europe Annual Symposium, *Science for health – beyond the silver anniversary*. Brussels, 24–25 March 2011.

Aggett, P., Antoine, J.-M., Asp, N.-G., Bellisle, F., Contor, L., Cummings, J., Howlett, J., Müller, D., Person, C., Pijls, L., Rechkemmer, G., Tuijtelaars, S. and Verhagen, H., 2005. PASSCLAIM – process for the assessment of scientific support for claims on foods. Consensus on criteria. *European Journal of Nutrition*, 44(Suppl. 1), pp. 1–30.

Barry, A., 2001. *Political machines. Governing a technological society*. London: Bloomsbury.

Biesalski, H.K., Aggett, P.J., Anton, R., Bernstein, P.S., Blumberg, J., Heaney, R.P., Henry, J., Nolan, J.M., Richardson, D.P., van Ommen, B., Witkamp, R.F., Rijkers,

G.T. and Zöllner, J., 2011. 26th Hohenheim Consensus Conference. Scientific substantiation of health claims: evidence-based nutrition. *Nutrition*, 27, pp. S1–S20.

Burchell, G., Gordon, C. and Miller, P., 1991. *The Foucault effect. Studies in governmentality*. Chicago, IL: The University of Chicago Press.

Callon, M., Lascoumes, P. and Barthe, Y., 2009. *Acting in an uncertain world*. Cambridge, MA, and London: MIT Press.

Callon, M., Méadel, C. and Rabeharisoa, V., 2002. The economy of qualities. *Economy and Society*, 31(2), pp. 194–217.

Chadwick, R., Henson, S., Moseley, B., Koenen, G., Liakopoulos, M., Midden, C., Palou, A., Rechkemmer, G., Schröder, D. and von Wright, A., 2010. *Functional foods*. Berlin: Springer, Coll. Wissenschaftsethik und Technikfolgenbeurteilung Band 20.

Coppens, P., Bijlsma, M., Craddock, N., Herreman, I., Hurt, E., Le Bail-Collet, Y. and Loosen, P., 2001. Are foods bearing health claims medicinal products? *Scandinavian Journal of Nutrition*, 45, pp. 140–145.

Dagognet, F. and Pignarre, P., 2005. *100 mots pour comprendre les médicaments. Comment on vous soigne*. Paris: Les Empêcheurs de penser en rond.

Diplock, A.T. *et al.*, 1999. Scientific concepts of functional foods in Europe – Consensus Document. *British Journal of Nutrition*, 81(1), pp. 1–27.

European Commission, 2003. Proposal for regulation of the European Parliament and of the Council on nutrition and health claims made on foods. COM(2003) 424 final.

Granger, E.H., 1984. Aristotle on genus and differentia. *Journal of the History of Philosophy*, 22(1), pp. 1–23.

Guéguen, D., 2007. *Lobbying Européen*. Brussels: Europolitique.

Guéguen, D. and Rosberg, C., 2004. *Comitology. The hidden power of the EU: finally a clear explanation*. Brussels: EIS.

Haraway, D., 1997. *Modest_Witness@Second_Millenium. Femaleman©_Meets_Oncomouse*. London: Routledge.

Heasman, M and Mellentin, J., 2001. *The functional foods revolution: healthy people, healthy profits?* London: Earthscan.

Hendrickx, K., 2013. Rivalling evidence-bases and politics in regulatory science. *Food Science & Law*, 4, pp. 125–128.

Hendrickx, K., 2014. *Bodies of evidence. An anthropology of the health claim*. PhD. Liège: University of Liège.

Howlett, J., 2008. *Functional foods. From science to health and claims*. Brussels: ILSI Europe Concise Monograph Series.

Irwin, A., Rothstein, H., Yearley, S. and McCarthy, E., 1997. Regulatory science – towards a sociological framework. *Futures*, 29(1), pp. 17–31.

Latour, B., 1997. Foreword: Stengers' shibboleth. In: I. Stengers, *Power and invention*. Minneapolis, MN: University of Minnesota Press.

Latour, B., 1999. *Politiques de la nature. Comment faire entrer les sciences en démocratie*. Paris: La Découverte.

Latour, B., 2004. How to talk about the body? *The Normative Dimension of Science Studies*, 10(2–3), pp. 205–229.

Latour, B., 2007. *L'espoir de Pandore. Pour une version réaliste de l'activité scientifique*. Paris: La Découverte.

Latour, B. and Woolgar, S., 1986. *Laboratory life. The construction of scientific facts*. Princeton, NJ: Princeton.

Marks, H., 1999. *La médecine des preuves. Histoire et anthropologie des essais cliniques 1900–1990*. Paris: Institut Synthélabo, Coll. Les Empêcheurs de penser en rond.

Marks, H., 2009. What does evidence do? Histories of therapeutic research. In: C. Bonah, C. Masutti, A. Rasmussen and J. Simon, eds. 2009. *Harmonizing drugs. Standards in 20th-century pharmaceutical history*. Paris: Editions Glyphe.

Mol, A., 2008. *The logic of care. Health and the problem of patient choice*. London: Routledge.

Pignarre, P., 1997. *Qu'est-ce qu'un médicament?* Paris: La Découverte.

Pignarre, P., 2004. *Le grand secret de l'industrie pharmaceutique*. Paris: La Découverte.

Pollan, M., 2008. *In defense of food. An eater's manifesto*. New York, NY: Penguin.

Roberfroid, M., 2000. Concepts and strategy of functional food science: the European perspective. *American Journal of Clinical Nutrition*, 71 (suppl.), pp. 1660–1664.

Scrinis, G., 2008. On the ideology of nutritionism. *Gastronomica*, Winter 2008.

Scrinis, G., 2013. *Nutritionism. The science and politics of dietary advice*. New York, NY: Columbia University Press.

Stengers, I., 2006. *La vierge et le neutrino*. Paris: Les Empêcheurs de penser en rond.

Stengers, I., 2011. *Cosmopolitics II* (R. Bononno, trans.). Minneapolis, MN: University of Minnesota Press.

Stengers, I., 2013. *Une autre science est possible! Manifeste pour un ralentissement des sciences*. Paris: La Découverte.

Timmermans, S. and Berg, M., 2003. *The gold standard: the challenge of evidence-based medicine and standardization in health care*. Philadelphia, PA: Temple University Press.

Troyer, R. and Markle, G.,1984. Coffee drinking: an emerging social problem? *Social Problems*, 31(4), pp. 403–416.

Young, A., 2005. The single market. A new approach to policy. In: H. Wallace, W. Wallace and M.A. Pollack, eds. 2005. *Policy-making in the European Union*. Oxford: Oxford University Press. pp. 93–112.

Part II

Shaping techno-natures

5 Competing, conflicting and contested futures

Temporal imaginaries in the GM crops controversy

Andreas Mitzschke

Introduction

Ideas about future technology scenarios offer powerful spaces for discussing and envisioning pathways for societal change. The enduring controversy about the potential risks and benefits of genetically modified (GM) crops is one of the arenas where different actors construct competing futuristic visions about the risks and benefits of technological innovation. As Giddens (1999) explains, controversies about technological risks revolve around the idea of controlling the future. In the GM crops debate, technological change is imagined to fundamentally shape society's future. Actors aim at shaping contemporary societal debate and policymaking with conflicting and contested imaginations. While imaginaries reflect the world as it is seen by the actors, they also facilitate thoughts about how the future ought to be. The competing imaginaries of the debate about the risks of GM crop technology open up pathways of thinking about agricultural futures while narrowing and excluding others.

This chapter employs the concept of 'sociotechnical imaginaries' (Jasanoff and Kim, 2009) in a textual analysis of advocacy material and qualitative interviews with key actors in the debate about the risks of GM crops in India and Europe.[1] Imaginaries are sociotechnical because science and technology are frequently imagined to generate social change and economic growth (e.g. Levidow and Papaioannou, 2013), of which the debate about GM crops is one example. Jasanoff and Kim emphasize that technoscientific imaginaries

> are not tied to future possibilities solely though scientific or technological practices. They are almost always imbued with implicit understandings of what is good or desirable in the social world writ large – for instance, how science and technology can meet public needs and who even are the relevant publics.
>
> (Jasanoff and Kim, 2009, pp. 122–123)

Such imaginaries of sociotechnical development provide different interpretations of the risks it entails and the public good it is thought to bring about. I argue that we can explain the competing and conflicting imaginaries by linking them back to the actor's different interpretations of the past and how these relate to the present. When we want to understand the controversy about GM crops through the lenses of imaginaries, we need to ask how we can understand these imaginaries in their temporality. In other words, how do they forge connections between the past, the present and the future? My aim is to show how various actors, by constructing futuristic visions with the help of science and technology, understand contemporary social reality and shape policy options for future agricultural development with or without GM crops. By tracing and characterizing the imaginaries in the controversy, I also want to understand where they come from, i.e. how they are actively maintained by reinventing and reappropriating the past.

My theoretical point of departure is the theory of social construction of technology (SCOT) (Pinch and Bijker, 1987; Bijker, 1995), which sociologically deconstructs the various meanings of a technological artefact: 'relevant social groups' share the same sets of meanings attached to a technology. In interaction between different actors in the controversy about GM crops, 'technological frames' structure the interaction between members of a social group. This is to show the 'interpretive flexibility' that describes how different relevant social groups interpret the same artefact in different ways.

I complement this theoretical frame by drawing on the constructivist and interpretive social science concept of 'sociotechnical imaginaries', which emphasizes the imaginative repertoires in the controversy. It is also conducive to studying the open-ended nature of the debate about GM crops, which has remained unresolved since the marketization of the technology in the 1990s. By accentuating certain arguments about the risks and benefits of GM crops through sociotechnical imaginaries, relevant social groups in the debate raise larger normative questions about agricultural futures, the definition of the public good, and various interpretations of past agricultural development. This addresses the co-construction (Bijker and Pinch, 2012; Jasanoff, 2004) of GM crop technology as the object of social and political concern with society and its publics. In terms of temporal imaginaries, past and future are co-constructed, i.e. how it is or was and how it ought to be are intertwined.

In tracing the sociotechnical imaginaries, my argument is that by constructing competing imaginations around GM crops, industry representatives, civil society organizations and scientists invoke conflicting normative visions of the future. Temporality is what connects these normative struggles: proponents and opponents in the debate place their conflicting views of the future with GM crops within different histories of agricultural development. The multiplicity of competing imaginaries are visions of the future which aim at shaping the present and yet are framed by various interpretations of the past – the construction of time acquires more prominence from such a perspective. For instance, while the critics of GM crops construct the technology as a continuation of industrial agriculture and, as such, a sociotechnical system with manifold hazards to humans and the ecosystem, proponents invoke an imaginary of linear technological development with unproblematic and manageable risks. While imaginaries are about the future, we also need to pay attention to how they connect to the past.

The following section explains how I mobilize the concept of 'sociotechnical imaginaries' for my analysis. What follows is a description of the sociotechnical imaginaries that I have identified for the proponents and opponents of GM crops in the debate in India and Europe. This is to show that the controversy is a normative struggle that finds expression in competing, conflicting and contested futures, yet these are maintained by diverging interpretations of the past.[2] I will then return to the aspect of temporality before I conclude with reflections on how to understand the debate about GM crops and its risks on the one hand, and conceptual implications about imaginaries on the other.

Sociotechnical imaginaries

The concept of sociotechnical imaginaries (hereafter referred to as STIs) was first used in anthropology to understand the social dimension of technological change (Marcus, 1995). Comparing the development and regulation of nuclear energy in the United States and South Korea, Jasanoff and Kim use the concept 'to show how different imaginations of social life and order are co-produced along with the goals, proprieties, benefits and risks of science and technology' (Jasanoff and Kim, 2009, p. 141). They show how the imaginary of nuclear energy is embedded in existing state–society relations and understandings of democracy through the notion of the public good, the construction of publics, and ideas about participation in science and technology decision-making. More recently, these authors have described STIs as 'collectively held and publicly performed visions of desirable futures, rooted in shared understandings of social life and order made possible by science and technology' (Jasanoff and Kim, 2015).

Within nation states, there are distinct cultures of imagining and implementing technological systems, and these differences are related to the nation's broader political development (Jasanoff and Kim, 2013). The idea of STIs is to show how national political cultures are intertwined with 'the development and reception of science and technology' (Jasanoff and Kim, 2009, p. 124). However, instead of trying to identify how imagination reflects a way of thinking about GM crops specific to the level of the nation state, I employ the concept of imaginaries differently. Relevant social groups in the debate about GM crops also construct a multiplicity of competing, conflicting and contested visions of the future that speak to and from different publics. Institutions, companies, environmental NGOs and expert bodies construct their own imaginaries, which operate within and between different contexts (Jasanoff, 2011; Jasanoff and Kim, 2015).

As the common-sense notion of imaginaries suggests, STIs are the ways actors conceive of a future with the help of science and technology. These imaginations are not exclusive to the latter, but they are closely intertwined with ideas of how the social world ought to be, e.g. by defining public goods, public needs and relevant publics themselves. In that sense, 'techno-scientific imaginaries are simultaneously also "social imaginaries", encoding collective visions of the good society' (Jasanoff and Kim, 2009, p. 123). The concept of STIs allows us to scrutinize continued contestation in debates that have not reached closure from a different perspective. By illuminating contested and competing dreams and ambitions shaped by the deep-seated background understandings of the actors involved, STIs allow us to reflect on competing normative conceptions of technological development that are co-produced with ideas about social life and order (Steger and James, 2013). By employing this concept to the debate about GM crops, I can show how imaginaries represent deeply held normative frames and belief systems.

Moreover, I argue that when studying competing STIs as a normative struggle we need to pay greater attention to their temporality. As such, the concept is future-oriented, but also constrained by present conditions and historical, political and cultural contingencies (Marcus, 1995), i.e. the concept relates various futures to the present and the past: relevant social groups link STIs of GM crops to ideas about the past, specifically globalization and the relationship between technology and environment in agriculture. By understanding how STIs are connected to such constructions of time, we can better understand how they translate and get repackaged into wider targets for public action and the construction of the very publics that are affected by GM crops.

Jasanoff and Kim allude to the scant attention paid so far to how STIs make and sustain the focus on some risks and benefits of sociotechnical change, while systematically downplaying other ideas about collective risk-taking (Jasanoff and Kim, 2013, p. 195). As Marcus suggests, 'the imaginary fills in the cognitive gap' (Marcus, 1995, p. 4) for analysing collectively held self-understandings of the social, technical and political. To understand the relationship between imaginaries and the discourse on risk as a way of social thought and practice, I draw on in-depth interviews with key actors in the European and Indian debates, information sources produced by non-state actors, and comments and reports in mass media. I follow the suggestion to study these sources in an interpretive textual analysis of the symbolic elements of such controversies (Jasanoff and Kim, 2009). As the following empirical investigation will show, individuals as well as groups imagine what the world might look like with science and technology; STIs thus make up a 'field of organized practice' (Appadurai, 1996) in which these actors' visions of the future become embedded in the temporality of social and normative structures.

The actors in the controversy about GM crops express their ideas of what is good and desirable, of the risks and hazards of GM crops. By focusing on the way that the risks and benefits of GM crops are represented through imaginaries, one can identify the imaginaries' conflicting definitions of the public good as well as their competing positions with respect to policy (Jasanoff and Kim, 2013). Presenting certain aspects of technoscientific development as most suitable or inevitable through futuristic visions makes STIs potentially 'hegemonic while seeming apolitical or value neutral' (Smith, 2009, p. 463). I therefore scrutinize the normative and prescriptive symbolic representation of STIs in the debate; I take imaginaries as a cultural resource that actors invoke (i) to understand technological risks, (ii) to shape policy preferences, and (iii) to construct publics relevant to such an endeavour (cf. Jasanoff and Kim, 2009, p. 122). I also trace how these imaginaries stabilize and consolidate through meanings attributed to the past. I argue that the actors' views about past agricultural development inform their perspectives about how GM functions within the global ecosystem and economy, and how they see the stakes and future implications behind certain policy options discussed in the debate.

The sociotechnical imaginaries of the GM crop debate

Productivity – food security and competitiveness

The proponents of GM crop technology construct GM crops within an imaginary of productivity. They present the technology to enhance yields to fight food shortages and world hunger more efficiently in the future. Hugh Grant, the CEO of the company Monsanto since 2003, explains in a 2008 television interview that GM crops reduce input costs and thus make farming more profitable. He presents the possibility for a better agricultural future with GM crops that respond to the need to increase yields to guarantee food security: 'Doubling food production in existing arable land with ever increasing demand for food, this objective cannot be realized without GM crop technology' (Grant, 2008). Those in favour of GM crops see the need to enhance agricultural productivity as a taken-for-granted reality. Productivity is an imaginary at work in India and Europe, though with different emphases in both locations.

In India, the imaginary of productivity for food security has a powerful cultural meaning that is rooted in the historical experience of food shortages and famine. Political leaders such as Pranhab Mukherjee (*Times*, 2013) or Union Minister for Agriculture Sharad Pawar (*Express*, 2013), biotechnologists (Padmanaban, 2013), and think tanks such as the National Academy of Agricultural Sciences (Sud, 2013) frequently invoke the idea that GM crops have a revolutionary potential to increase agricultural productivity. In the face of this vision, the proponents of GM crops consider regulation of the technology to be too strict (Rao, 2013a) and regard the risks of GM crops as negligible compared to GM's potential benefits: 'Why are those risks? Any of them has no meaning at all, absolutely there is no meaning, because look at malnutrition, look at starvation, all happening in this country' (Interview, Padmanaban, 2012).

While productivity for food security appears to be a powerful imaginary in India, in Europe it primarily serves as a means of rhetorically promoting GM crops as an altruistic technology for development (Glover, 2008, 2010). GM advocacy organizations, such as EuropaBio, the European association for biotechnology industries, presents GM crops as a panacea for the problem of food shortages: 'We need to produce more food, we need to produce it faster and more efficiently' (Interview, Servaas du Marchie, 2012).[3] More importantly, however, they present GM crops as a necessary innovation for the European farming sector's competitiveness (EuropaBio, 2014b), as a technological innovation that should be made available for the farming sector (EuropaBio, 2011) in order to reach substantial economic gains (EuropaBio, n.d.-a):

> Continued non-availability of the technology for EU farmers essentially leads to a competitive disadvantage compared to global competitors on world markets. As new seeds continue to come to the market in other countries, but not the EU, this gap will widen.
>
> (EuropaBio, 2011, p. 12)

The problem definition in this imaginary is the taken-for-granted assumption that hunger and malnutrition need to be fought by increasing agricultural productivity. The risk is one of missing a technological opportunity for improving the social condition, either in terms of food security or in regards to competitiveness. Genetic modification is thus constructed as the inevitable panacea to increase yields and to contribute to economic growth. The audiences this imaginary refers to are not actively participating in the discourse but are rather presented as passive recipients of the public good that GM crops are imagined to entail – enough food for all, and gaining a competitive edge in global agricultural markets. The framing of GM crops as a concept of food security and global competitiveness constitutes the technology as a moral imperative in itself.

How is the imaginary of productivity rooted in the past? The proponents of GM crops see the technology as a precise intervention not much different from but complementary to the agricultural technologies of the Green Revolution. The Green Revolution refers to the period between the 1940s and 1960s in which agricultural production was boosted worldwide by technological innovations such as high-yielding hybrid seeds, the expansion of irrigation facilities, the application of inorganic fertilizers and the use of chemical pesticides.[4] Biotechnologists describe GM as a technique that is more accurate and quicker than conventional breeding. Producers and scientists working on GM crops imply that risks are due to bad management practices, and that they can be coped with by further refining crop traits with genetic modification (Gheysen, n.d.; Monsanto, 2012; Rao, 2013a). The imaginary of productivity overlaps with discourses of the Green Revolution that extend instrumental reason to agriculture (Visvanathan, 1997, p. 74). As Books points out, the Green Revolution reverberates in the discourse about GM crops in presenting food shortages as a supply problem. The imaginary of productivity envisions complex socio-economic realities to be controlled by having more science in agriculture, by adding technological fixes, by intensification of agricultural practices and by the transformation of the field into monocultures (Brooks, 2005).

Environmentalism

The proponents of GM crops also construct the imaginary of genetic modification as an environmentally sustainable technology. A European industry representative remarks: ‘GM crops have the potential to increase GDP growth while reducing pesticide applications at the same time’ (Interview, Servaas du Marchie, 2012). The proponents see the reduction of pesticide use in agriculture as benefiting the environment, an issue that has traditionally been the concern of environmental NGOs and civil society organizations. However, as studies on agricultural multinational corporations suggest, 'environmentalism' is a malleable concept, a set of ideas and guiding principles that corporations try to mobilize to convince buyers that their products are more environmentally friendly in order to sustain and enlarge market shares of their products (Jansen, 2004; Jansen and Vellema, 2004).

The sociotechnical imaginary of environmentalism is at work in India and Europe, too. An Indian industry representative contends, 'there are social advantages because there are fewer [pesticide] sprays. So, environmentally it is safer ... the farmers will benefit' (Interview Setharama, 2013). In this vein, GM crops are constructed as an option for sustainable agriculture: 'Bt crops are the best option for organic farming. The function of Bt is to decrease pesticide use' (Interview, Padmanaban, 2012).[5] As a member of the Association of Biotech-Led Enterprises (ABLE) suggests, picking up on environmental issues and constructing the possibility of future technological wonders around them will lead to greater public acceptance of the technology: 'bio-tech has endless possibilities. And those kinds of technology will be more acceptable to the public' (Interview, Setharama, 2013).

Advocacy organizations in Europe are equally active in constructing GM crops as sustainable. The argument is that the technology can reduce agricultural impact on biodiversity when used with sustainable crop management techniques. They present the benefits of insect resistance (IR) traits as reducing pesticide spraying, and that herbicide-tolerant (HT) GM crops will lead to less carbon output and protect soils from erosion and compaction through less ploughing (ABC, 2012; EuropaBio, n.d.-b). GM crop producer Monsanto Company claims that since the introduction of IR GM crops, such as Bollgard cotton, the application of pesticide sprays has been reduced by 18 per cent; however, they do so without specifying scientific sources or the time frame of measurement (Monsanto, 2012, p. 10, p. 126). Published papers on the environmental impact of GM crops written by biotechnology consultancy organizations present data to support such claims (e.g. Brookes and Barfoot, 2011; Barfoot and Brookes, 2014).

The issue of pesticide applications allows for a discursive combination of environmentalism with economic benefits. The use of agricultural pesticides is presented as a risk in itself that can be mitigated by GM crops. This serves as a powerful imaginary that presents the farming community and society at large as the benefactors of GM crop technology. Environmental sustainability is therefore constructed as a public need to which the imaginary of environmentalism responds. Most of the current globally operating GM seed producers entered the crop biotechnology market when their core business in the production of chemical agricultural inputs such as pesticides began to show signs of decreasing growth in the early 1980s. At the same time, this was a moment when the environmental movement had grown to maturity and successfully invoked sympathy for their arguments about the environmental consequences of input-intensive farming among larger parts of the public (Schurman and Munro, 2010).

The critique of the Green Revolution constitutes a reference point around which actors in favour of GM crops can connect. Several interviewees argued that GM crops respond to a history of environmental concerns, such as the move from first- to second-generation GM crops (e.g. Interview, Servaas du Marchie, 2012).[6] Constructing science as a continuum that provides innovative technologies to deal with problems that arise in technological systems, a philosophical perspective known as the idea of the 'techno-fix' (Weinberg, 1966, 1991), allows the proponents of GM crops to construct the technology as progress vis-à-vis the memory of environmental degradation associated with the Green Revolution. As an Indian industry representative contends: 'in fact, the Green Revolution is the culprit, not GM technology' (Interview, Setharama, 2013).

Still, the positive historical memory of the Green Revolution period remains the dominant interpretation of the past, projected onto the future within both the STIs of productivity and environmentalism. As an Indian biotechnology advocate explains, 'the Green Revolution demonstrated that various barriers to food production could be overcome through technological intervention' (Rao, 2013a, p. 167, p. 171). Rao presents agricultural biotechnology as an extension of previous technological improvements in agriculture. To him, GM crops are a logical continuation of the technological culture of the Green Revolution, which assumes a linear causality of rising yields to improve the human condition from hunger to economic growth. From this view, GM crops constitute an improved breeding technology, with greater precision and speed. Emphasizing the continuity of technological progress in effect naturalizes GM crops as the necessary outcome of a linear historical development of agricultural technologies that has been yielding ever greater efficiency and enhanced precision:

> All the ten thousand years history of agriculture … is genetic modification only. It was done initially by selecting natural mutations and then you induced mutations: first natural hybridization … and later artificial hybridization. In artificial hybridization, you are putting two entire genomes. In mutation, it is hit and miss. You just don't know what is happening.… You never know how many bad genes have been fixed in that. In that way, genetic engineering is more precise.… They are eliminating anything that is not acceptable.
>
> (Interview Rao, 2013d)

Actors in opposition to GM crop technology, from Indian activists to European transnational NGOs, do not buy into the lifeworld and temporality of these STIs. Instead, they question their epistemic foundations (e.g. Kuruganti, 2009; Zacune, 2011). To them, the logic of this sort of environmentalism is rooted in the imaginary of productivity, too, since the STI of environmentalism poses a chimera: 'the camouflaged description of transgenic crops hides many of the ecological impacts of genetically engineered crops. The illusion of sustainability is manufactured' (Shiva, Barker and Lockhart, 2011, p. 12). In constructing entirely different visons of the future and what is desirable for society, the opponents of

GM crops invoke STIs based on different sources of knowledge and different normative claims. Instead of taking for granted the positive visions about the alleged benefits of GM crops, they construct the risks of this technology in a profoundly different way: GM crops are part of a horrific future in which the ecosystem is out of balance and where farmers have lost control over their choice of seeds.

Complex ecology

Indian environmental activist and writer Vandana Shiva is one of the most prominent spokespersons for the imaginary of complex ecology.[7] She imagines a disastrous future with GM crops:

> if that system was allowed to spread.... A destroyed planet will give no food at all. Dead soils, disappearing waters, a totally chaotic climate, no seeds – it is a recipe for an absolute, not just disaster, but a recipe for human extinction.
>
> (Shiva, 2014)

This dystopian idea relates back to a notion of GM crops as an unpredictable large-scale intervention into the ecosystem, and the lack of knowledge and scientific certainty regarding the complex interdependencies therein. Social groups critical of the technology draw on complex ecosystem interactions in terms of insect and weed resistance, secondary pests and the impact on non-target organisms in constructing this STI.

Envisioning the worst future with GM crops, a report by an international coalition of civil society organizations suggests that 'instead of controlling weeds and pests, GM crops have led to the emergence of super weeds and super pests' (Shiva, Barker and Lockhart, 2011, p. 12). Their report gives the example of pollination between cultivated GM sugar beets and weed beets that might undermine the efficacy of HT traits through outcrossing and amplification of the relevant genes into wild species (cf. Boudry *et al.*, 1993; Powles, 2010). Due to the development of resistance in weeds, the report foresees an increase in the overall use of herbicides. To take home the point, it draws on a study commissioned by the US Environmental Protection Agency that confirms the use of herbicide-tolerant crops in the United States has led to a tenfold increase of pesticide use in the period 1993 to 2007 (Grube *et al.*, 2011).

This STI foresees an apparent risk that pests will develop a resistance to the Bt toxin: 'The question is not whether super-pests will be created, but when they will become dominant' (Shiva, Barker and Lockhart, 2011).[8] Anti-GM actors in Europe make the same argument. They envision a future in which farmers have to resort to more intensive use of pesticides due to the development of resistance in weeds and insects (Zacune, 2011; Antoniou, Robinson and Fagan, 2012, 2014; Greenpeace Nederland, 2012). A Dutch Greenpeace activist remarks with reference to HT crops that the ecosystem will adapt more quickly than scientists can develop GM crop technology in the laboratory: 'HT crops pose specific environmental problems. Weeds develop a resistance to the pesticide that is constantly used. That is how nature, or evolution works' (Interview, van Bekkem, 2012). In reference to IR crops, this interviewee constructs the genetic modification of crops as a technology that is too simplistic vis-à-vis the complexity of the ecosystem:

> If you succeed in fighting one insect with a Bt toxin, the ecological niche of this specific organism is taken over by a secondary pest, which maybe is not vulnerable to the Bt toxin. So you still have to spray pesticides. And the insects which are dying from the Bt toxins, some will survive and the population will develop resistance. In the end the farmer sprays more and more chemical pesticides.
>
> (Interview, van Bekkem, 2012)

The opponents also argue that, even if the Bt toxin works on the target pest, secondary pests fill in the ecological niche, in response to which farmers will have to resort to pesticides: 'even if successful at controlling a target pest species, other pests (called "secondary pests") may then emerge as more prominent threats to the plants, resulting in crop loss and the need to apply additional pesticides' (Hammond, 2010, p. 2). Activists in both India and Europe argue that ecological complexity is difficult to control with very specific agricultural technologies such as IR crops; i.e. the GM trait is too specific, and thus carries an intrinsic risk (Antoniou, Robinson and Fagan, 2012, 2014).[9]

GM crops themselves pose a risk to the stability of the ecosystem, rather than a solution to the risks of agricultural practice. The technology is interpreted to be a continuation of input-intensive chemical farming: GM crops endanger ecosystem diversity because of their manifold impacts on organisms in and around agricultural fields. This point is illustrated by the question of the impact of the existing GM crop traits on non-target organisms (NTOs).[10] Shiva, for example, states that:

> the Bt crop does not affect 'just one specific pest'. Beneficial insects like bees and ladybirds can be seriously affected. A Cornell study showed that the Bt toxin affected the Monarch butterfly. Navdanya's studies have shown that soil micro-organisms are negatively affected.
>
> (Shiva, Barker and Lockhart, 2011, pp. 12–13)

Aside from obvious calls on stricter environmental risk assessment, the policy implications that the imaginary of complex ecology calls for is to pair knowledge and expertise of agro-ecology with its privileged counterpart, agricultural economics, as the dominant epistemic frame for evaluating agricultural technologies. As Indian activist Suman Sahai explains, GM is being presented as the unavoidable route to the future, whereas agro-ecological expertise stays marginalized:

> We have fine scientist who are plant breeders, soil scientists, etymologists, agronomists – who are doing very good work. Unfortunately, in this completely mad race for GM products, these sectors of agricultural science … get neglected in favour of this chimera called GM technology.
>
> (Interview, Sahai, 2012)[11]

The risks can be subsumed as being about the potential consequences for the stability of complex ecosystems, the latter thus arising as a public good. Opposed to this is the simplicity and linearity of GM crop technology based on the reductionism of the epistemic foundations of modern input-intensive agriculture. The STI of complex ecology clearly relates visions of possible futures with and without GM crops to the interpretation of the Green Revolution as a violent intervention into social and natural worlds, yet yielding heavy costs. A long-time Indian activist and organizer of a national anti-GM network contends: 'now, it is crossroads time, we are at the point to make decisions based on the last 50 years of experience in agriculture' (Interview, Kuruganti, 2012).[12] This view has been developed by Shiva (Shiva, 1991), as well as in the seminal book *Silent Spring* (Carson, 1962), to which activists frequently refer when talking about GM crops.[13] However, GM opponents also construct a positive imaginary in which agro-ecology plays a decisive role for agricultural change, thus breaking with the past. They see the global South providing a fertile ground for an alternative imaginary of the future: 'people of Europe and America, I think they are lacking a dream. We need a dream.… People should have an idea what could be different' (Interview, Debjeet, 2012).

Seed sovereignty

The opponents of GM crops also invoke the STI of seed sovereignty, from which they criticize GM crops as being part of the project of corporate-driven neoliberal globalization. To them, GM crops pave the way to monopolistic markets and to the loss of public control over seeds. This revolves around the idea of choice attached to farming practice and its inputs. Often, arguments coming from the STI of complex ecology provide an entry point to the slightly more complicated notion of seed sovereignty:

> The fable that GMOs are feeding the world has already led to large-scale destruction of biodiversity and farmers' livelihoods. It is threatening the very basis of our freedom to know what we eat and to choose what we eat. Our biodiversity and our seed freedom are in peril. Our food freedom, food democracy and food sovereignty are at stake.
>
> (Shiva, Barker and Lockhart, 2011, p. 7)

Here, activists link GM crops as an issue of agriculture with notions of democracy. They explain their fear of a future with GM crops as one in which multinational corporations reduce the choices for cultivating seeds by establishing seed monopolies.[14] A Friends of the Earth Europe spokesperson explains the issue of seed sovereignty as a global concern to the farming community:

> The idea that a few multinationals control what happens to seeds is a nightmare for everyone with an understanding of agriculture. The local context might be different, but everyone seems to agree on this point, independent of where in the world they work in agriculture: If you cannot decide yourself what happens to seeds … [t]here is the fear that food production will be taken out of the hands of the farmers.
>
> (Interview, Schimpf, 2012)

Indian opponents link the question of GM crops posing a risk to sovereignty much more explicitly to the critique of globalization. Their discourse is framed by a postcolonial understanding of the global power relations they see inscribed in technoscientific projects for the modernization of agriculture, e.g. through the Green Revolution. The future they imagine with GM crops is one where sovereignty is at risk because diversity is reduced by seed technologies that favour large-scale monocultures and economies of scale. A farmer union leader explains that, in contrast, 'agriculture depends on diversity, not technological interventions' (Interview, Singh, 2013).

In the imaginary of seed sovereignty, activists construct risk as the loss of farmers' control over productive resources, starting with the availability and choice of seeds. As an organic farming activist explains, GM crops instead destroy alternative imaginations by reducing options for an agricultural future not determined by market principles: 'The farmer has been transformed into a consumer of agricultural inputs and this is rooted in the transition towards modern society' (Interview, Babu, 2013).[15] This critique explicitly targets a model of globalization that disproportionately benefits the interests of multinational corporations through the privatization of agricultural inputs in developing countries and the associated disruptions in the social organization of rural labour. Therefore, these actors direct their critique not only at the technology itself, but also at policy and market environments that surround it:

> Now there is complete control [of the seed market]. There is no small seed store left in the US or in Europe.… they imposed this model on the rest of the world through agencies like WTO and its agreement on agriculture, etc. They were not satisfied with just the European market; they wanted to look at the global market. And how do you do that? You want policy changes in those countries where you want the market. That is why we have seed patents through TRIPS and the WTO.
>
> (Interview, Sahai, 2012)

Against this background of a fundamental critique of the institutions of globalization, activists see the promotion of agro-ecological alternatives as a policy imperative, yet marginalized in the face of the predominant system of industrial agriculture:

> Alternative models in terms of technologies, institutions, policies and markets ... these alternatives are not seen as fitting scaling up attempts because they do not fit the current institutional system in agriculture.... Diversity is key in technology and the political system.
>
> (Interview, Ramanjaneyulu, 2013)

In that sense, GM crops become an issue of freedom and democracy, rather than the threat of a seed market dominated by multinational biotech corporations and the rationale of industrialization. As Saldanha, an activist from Environment Support Group (ESG),[16] explains, 'a pluralistic view of this technology is not just about human health and environment it is also about seed sovereignty' (Interview, 2012). This line of argumentation is made explicit by constructing the farming community as the debate's central audience:

> The principal players are the farming groups, the organic farmers, the independent experts, a variety of civil society organizations. For me that is a positive thing. I have always been wary about corporatized NGOs coming into India and clouding and conforming us. And I have always argued that is not our ethos. Our whole philosophy is that we should not become corporatized. We should become supportive of a deeper democratization, of debate and these sorts of things.
>
> (Interview, Saldanha, 2012)

How is the STI of seed sovereignty related to the past? Reflecting on the history of agricultural transformations, GM crops are seen as a continuation of a historical legacy. The activist suggests that after colonialism the dominance of economic elites continues through the process of globalization. In his view, the project of GM crops fits into the structure of power relations that control industrializing and developing nations through agricultural technologies: 'The structures are the same. Capitalism has adopted the colonial structures. It is imperialism. What is Monsanto but a wing of the United States?' (Interview, Saldanha, 2012). Another interviewee facilitates the explanation of GM crops as a continuation of colonial power relations: 'GM is certainly an extension of the past. It is the same set of players who want to make money with the old and new technology' (Interview, Kuruganti, 2013). The past of colonialism serves as a backdrop for understanding current developments in agriculture from a wider historical scale stretching over longer periods, and therefore for an imaginary of the future.

Time and temporality

Each of the four sociotechnical imaginaries outlined above suggests a different perspective on the normative implications of GM crop technology and to the temporal relations associated with it – a divergent field of conflicting and competing normative and temporal relations that make up entire world views. The STI of productivity constructs agriculture as implicated with the problem of how to increase yields while the STI of environmentalism picks up the idea that GM crops can solve environmental problems. Both perspectives suggest a linearity of technological fixes to alleviate human and environmental problems informed by market imperatives as a guiding principle. This worldview clashes with the STI of complex ecology that sees the fragile interdependency of the ecosystem as being at risk and the STI of seed sovereignty in which farmers lose the sovereignty over available seeds, so essential for retaining autonomy over the social relations of production in farming. Inspired by discussions on scale in the social sciences, I suggest that a perspective on temporality can help us better grasp these conflicting and competing imaginaries, though this methodological idea can only be presented in its infancy here and should be taken as a suggestion for further research.

Adam (2006) suggests that the focus on time allows for a different historical perspective, and Sharma (2013) reminds us that studying temporalities opens up spaces for so far unacknowledged political and normative implications. My investigation of the temporal relations of the STIs in the GM crop debate indicates that different perspectives of time inform the normativity of actors' arguments; how they relate to the past shapes the construction of their imaginaries. Adams infers that the naturalized machine time of the West now spans the globe and imposes a time frame marked by invariability, standardization, quantification and universalization. The STIs of productivity and environmentalism embody these dimensions in accounting for benefits in terms of yield and in reacting to environmental concerns by means of universal techno-fixes. Although naturalized machine time has benefited the global span of capitalist and industrial development, it fails to take into account the temporality of socio-economic and ecological time frames, thus leaving 'other temporal orientations and structures de-legitimised' (Adam, 2006, p. 124).

The STIs of complex ecology and seed sovereignty show how actors try to engage with an alternative to the temporal imposition of industrialization and globalization as processes that colonize both nature and society. While proponents of GM envision the technology as part of an invariable course of human and technological progress, the STIs of complex ecology and seed sovereignty suggest different temporal frames. By developing counter-narratives on the longer timespans of ecosystem development and colonial relations, these STIs try to shift emphasis onto the *longue durée* of socio-historical time. Historian Fernand Braudel has discussed how society stretches over short and long periods of time and emphasized how not only material but also mental and environmental structures shape the long-term course of events in human history beyond the immediate consciousness of the actors involved (Hall, 1980). The idea that reverberates with the STIs of complex ecology and seed sovereignty is that ecological, social and cultural structures evolve over considerably different and longer time frames than technoscientific development does. STIs offer attractive mental frames that correspond to normative ideas about technological and social development. A temporal perspective on STIs therefore allows us to understand imaginaries of technological development as implicated by converging and clashing normative perspectives, which allows an inquiry into 'time politics' (Sharma, 2013).

Taking such a perspective into account allows us then to see the normativity of the respective STIs vis-à-vis other perspectives within the history of sociocultural development – the actors' deep-seated background understandings and interpretations of the past inform their imaginaries of the future. The resulting STIs as a construction of a possible future with science and technology are embedded in mental and normative structures and these in turn relate back to different perspectives of how the longer time frames of the ecosystem, industrialization or global capitalism evolve. Conceptually, this implies that understanding STIs from a temporal perspective allows us to take into account actors' complex attachments, e.g. their relationship to ecology, farming and globalization. The notion of uniformity and universality that underlies the normative arguments of the STIs of productivity and environmentalism stands in contrast to the norms of diversity and interconnectedness underlying the STIs of complex ecology and seed sovereignty. Visvanathan (1997) explains that the concept of diversity in the farming context means more than ecological or genetic variety. It goes beyond that to represent the multidimensionality of natural processes and thus defies the universalizing reductionism and economic rationalization ascribed to technological globalization. In that sense, pointing to the underlying norms of the STIs identified above – and their respective time frames – opens up the space for analysts to describe the politicization of time and the contours of power that the temporalities of sociotechnical imaginaries in the GM crop debate entail (cf. Sharma, 2013).

Conclusions

This chapter began by asking how we can understand the sociotechnical imaginaries in the GM crops debate vis-à-vis their temporality, i.e. how actors relate interpretations of the past, the present and their visions of the future. The concept of sociotechnical imaginaries (STIs) provides a different look at the controversy by highlighting the imaginative repertoires of the actors. This allowed me to show the larger normative questions about technology and society, the shaping of nature through agriculture, and more generally, the relationship between technological development and visions of social order at play in the controversy. In that sense, STIs are a form of value-laden, technologically enabled visions of the future.

This chapter charted how sociotechnical imaginaries of productivity, environmentalism, complex ecology and seed sovereignty conflict and compete. These imaginaries stem from different normatively founded conceptions of risk, policy imperatives and definitions of the public good. I argue that we can better understand these contested imaginaries by linking them back to the actors' various interpretations of past agricultural development and ideas about time – temporality is what connects these normative struggles. When actors refer to wider processes of environmental and sociocultural change such as ecosystem evolution and globalization, there are different historical temporalities at play: economic, ecological and political. Studying these through the lens of temporal sociotechnical imaginaries invests the concept with a critical perspective on the role of time (cf. Sharma, 2013).

The STIs of productivity and environmentalism construct the future of agricultural development through science and technology as one of linear progress through time, starting with the Green Revolution. In contrast, the STIs of complex ecology and seed sovereignty emphasize larger timescales and processes such as colonialism, globalization and ecology. Temporality is what connects notions of the complexity, indeterminacy and interconnectedness of nature and its technical mastery through agriculture to understandings of sociopolitical and ecological processes. Questions about the colonial legacy of contemporary globalization and the interdependency of ecological processes open up normative considerations to questions of global power structures. Causalistic reductionism opposes the complexity of nature, and the dominance of market imperatives contrasts with the notion of autonomy and independence. Each of these perspectives is rooted in imaginaries and their temporality, assumptions about the past that shape the current lifeworld of the actors and influence their visions of a future with GM crops. Imaginaries thus tempt their audiences to buy into the mental structure of a whole lifeworld connected by normativity and temporality.

STIs as a concept allow us to connect the mental and the material and thus offers a different perspective on the co-construction of technoscientific futures and ideas about the past and the future. The latter is equally imagined, interpreted and appropriated by the actors in the debate. Although imaginaries get shaped locally, the visions of social order they construct entail deeper normative and temporal notions of how society ought to fit together and what role science and technology take therein (Jasanoff and Kim, 2015). A temporal perspective of sociotechnical imaginaries allows us to grasp the dimensions of debates about risk that have so far gone untouched. The methodological implications I have charted out above are only in their infancy and need to be explored beyond the study of the GM crop debate.

Notes

1 Qualitative, semi-structured interviews were conducted in the years 2012–2013 in various places in India and Europe. Textual sources were largely derived from online sources and print material and consist of civil society reports, industry information material, reports and speeches by leading figures in the debate.

2 The focus here is not on a systematic comparison between India and Europe. The empirical data for this chapter stem from the multi-sited ethnography in both regions for the author's dissertation research, which addresses the issue of comparison more explicitly. For the purposes of this chapter, India and Europe are both interesting because of their commonalities in public opposition to GM (food) crops and the difficulty for GM crop commercial products to enter the market for cultivation.

3 Conceived as the voice of the biotech industry in Europe, EuropaBio lobbies the relevant institutions of the European Union for a more favourable business environment for biotechnology companies (Paul *et al.*, 2004). The website Powerbase, a database by civil society organizations critical of corporate influence on policymaking, describes EuropaBio as a lobby organization that aims to win public trust in the safety of GM crops (Powerbase, 2010). Based in Brussels and consisting of corporate members and national biotechnology associations, EuropaBio represents more than 1800 companies, among them major transnational corporations like Monsanto, DowAgro Sciences and Bayer CropScience (EuropaBio, 2014a).

4 For continuities between the Green Revolution and the 'Gene Revolution', a term ascribed to GM crops, see Brooks (2005).

5 Bt refers to the soil bacterium Bacillus Thuringiensis, a genetic sequence of which has been incorporated in a wide range of crops, such as maize and cotton. This gene transfer makes the plant produce a toxin that kills insects feeding on it, thus making it insect resistant (IR). The argument is that by making the plant produce a toxin, production becomes more efficient and cost-effective by reducing the need to apply chemical pesticides externally and by increasing yields. Field-level data suggest that this might be true (e.g. Kathage and Qaim, 2012; Krishna and Qaim, 2012), while a growing body of literature questions such positive interpretations of the existing data (e.g. Dorhout and Rice, 2010; Ranjith, Prabhuraj and Srinivasa, 2010; Gassmann *et al.*, 2014).

6 First-generation GM crops were engineered to express a single trait such as herbicide tolerance or insect resistance in a single variety. Combining multiple of these traits in one crop variety is referred to as second-generation GM crops. The third generation, which has not yet left the laboratory and greenhouse stages of development, promises entirely new traits such as improved nutritional value and resistance to disease, as well as saline and drought resistance for GM crops.

7 Shiva has published extensively on livelihood and ecology (Shiva, 1988), the Green Revolution (Shiva, 1991) and GM crops (Shiva, 2000). While some describe her as a rock star of the global fight against agricultural biotechnology (Specter, 2014), she also receives criticism for frequently exaggerating her claims about GM crops (Herring, 2006). Many Indian grass-roots activists question the legitimacy of her campaign as representative of India's civil society and farming community, while internationally, and particularly in the West, she is represented as 'the voice leading the crusade against GMOs' (Frankman and Weinberger, 2014). Although Shiva might not represent the entirety of the anti-GM movement, she can nevertheless be considered to be an influential spokesperson providing widely perceived narratives, arguments and imaginaries about GM crops.

8 The US Environmental Protection Agency recommends non-Bt crop refugia next to Bt fields to manage pest resistance. Indeed, as policymakers and scientists are aware, pest resistance is unavoidable and can only be delayed with careful crop management, crop rotation and resistance management practices (cf. EFSA, 2011; Gassmann et al., 2011). Whether in diverse agrarian systems as that of India, such complex farm level management practices centred around GM crop technology alone are feasible remains open to debate. Studies of integrated pest management indicate that a larger variety of options is most practical in the Indian agricultural context (Quartz, 2011).

9 The opponents of GM crops draw extensively on scientific studies, e.g. on field-level studies from China, India and the United States (e.g. Dorhout and Rice, 2010; Zhao, Ho and Azadi, 2010) to give their arguments credibility.

10 NTOs are living organisms that are present in the field but which do not reduce yield. Non-target organisms can include soil organisms, non-pest insects, birds and other animals. Conventional agricultural methods (herbicide and pesticide application in particular) have large and well-documented non-target effects. While GM crops may have more specific target effects, non-target effects cannot be principally ruled out

(Pilson and Prendeville, 2004; Snow et al., 2005), particularly if pesticide use is not decreasing as expected in the long run.

11 Dr Suman Sahai has a PhD from the Indian Agricultural Research Institute. She has worked as a geneticist and founded the civil society organization Gene Campaign in 1993. The organization describes itself as active in the field of research and advocacy for equitable access to and use of bio-resources (GeneCampaign, 2014).

12 Kavaitha Kuruganti is the convenor of the Alliance for Sustainable and Holistic Agriculture (ASHA), an India-wide informal network that concerns itself with bringing together actors and knowledge about 'food, farmers and freedom' (ASHA, 2010). Kuruganti has been an activist for more than two decades; she writes a blog (Kuruganti, 2015) and frequently contributes to debates about agriculture and GM seeds in national news.

13 In her book, Carson describes the impact on the environment and human body of unregulated use of chemical pesticides. She depicts pesticides as a technology of conquest to dominate and shape nature and opposes this with a view on ecological complexity that pays attention to the long-term consequences of technology.

14 Environmental and farmers' movements allude that the introduction of GM crops 'has corresponded with increasing monopolization of seed by biotechnology companies and higher seed costs' (Roseboro, 2013). Studies of the GM hybrid seed market for cotton in India suggest monopolistic tendencies since the introduction of the technology (Murugkar, Ramaswami and Shelar, 2007).

15 Babu is a grass-roots activist who works for the Institute for Cultural Research and Action in Bangalore, which engages in action research to promote organic agriculture. It works directly with farmers and explicitly criticizes the dominant paradigm of development (ICRA, n.d., Interview P. Babu, 2013).

16 ESG is a non-profit NGO placed in Bangalore. It works with various Indian groups dedicated to issues of environment and social justice to support research, education and campaigns on environmental issues. ESG describes its ethos as 'to be sensitive to contextual complexities so the solutions are environmentally and social just, and deeply democratic' (ESG, n.d.).

References

ABC, 2012. *Going for growth*. Agricultural Biotechnology Council (ABC).

Adam, B., 2006. Time. *Theory, Culture, Society*, 23(2–3), pp. 119–138.

Antoniou, M., Robinson, C. and Fagan, J., 2012. *GMO myths and truths. An evidence-based examination of the claims made for the safety and efficacy of genetically modified crops*. London: Earth Open Source.

Antoniou, M., Robinson, C. and Fagan, J., 2014. *GMO myths and truths. An evidence-based examination of the claims made for the safety and efficacy of genetically modified crops*. 2nd ed. London: Earth Open Source.

Appadurai, A., 1996. *Modernity at large. Cultural dimensions of globalisation*. Minneapolis, MN: University of Minnesota Press.

ASHA, 2010. *Kisan Swaraj Yatra – towards a Kisan Swaraj policy*. [Online] Alliance for Sustainable and Holistic Agriculture (ASHA). New Delhi. Available at: www.kisanswaraj.in [accessed 6 June 2016].

Babu, P., 2013. Interview, Institute for Cultural Research and Action (ICRA), Bangalore, 11 January 2013.

Barfoot, P. and Brookes, G., 2014. Key global environmental impacts of genetically modified (GM) crop use 1996–2012. *GM Crops & Food: Biotechnology in Agriculture and the Food Chain*, 5(2), pp. 149–160. doi:10.4161/gmcr.28449.

Bijker, W.E., 1995. *Of bicycles, bakelites, and bulbs: toward a theory of sociotechnical change*. Cambridge, MA: MIT Press.

Bijker, W.E. and Pinch, T., 2012. Preface to the anniversary edition. In: W.E. Bijker, T.P. Hughes and T. Pinch, eds. 2012. *The social construction of technological systems. New directions in the sociology of technology*. Cambridge, MA, and London: MIT Press. pp. xi–xxxiv.

Boudry, P., Morchen, M., Saumitoulaprade, P., Vernet, P. and Vandijk, H., 1993. The origin and evolution of weed beets – consequences for the breeding and release of herbicide-resistant transgenic sugar-beets. *Theoretical and Applied Genetics*, 87(4), pp. 471–478.

Brookes, G. and Barfoot, P., 2011. *GM crops: global socio-economic and environmental impacts 1996–2009*. Dorchester: PG Economics.

Brooks, S., 2005. Biotechnology and the politics of truth: from the green revolution to an evergreen revolution. *Sociologia Ruralis*, 45(4), pp. 360–379.

Carson, R., 1962. *Silent spring*. Boston, MA: Houghton Mifflin.

Debjeet, 2012. Activist, Living Farms Bhubaneswar, Bhubaneswar, 17 March 2012.

Dorhout, D.L. and Rice, M.E., 2010. Intraguild competition and enhanced survival of Western beat cutoworm (lepidoptera noctuidae) on transgenic Cry1Ab (MON 810) bacillus thuringiensis corn. *Journal of Economic Entomology*, 103, pp. 54–62.

EFSA, 2011. Scientific opinion on application (EFSA-GMO-CZ-2008–54) for placing on the market of genetically modified insect resistant and herbicide tolerant maize MON 88017 for cultivation under Regulation (EC) No 1829/2003 from Monsanto. *EFSA Journal*, 9(11), p. 2428.

ESG, n.d. *Environment support group – trust: about us*. [Online] available at: www.esgindia.org/about-us/index.html [accessed 24 December 2014].

EuropaBio, 2011. *GM crops: reaping the benefits, but not in Europe*. 2nd ed. Brussels: EuropaBio.

EuropaBio, 2014a. *About EuropaBio*. [Online] available at: www.europabio.org/about-europabio [accessed 2 September 2014].

EuropaBio, 2014b. Biotechnology industry manifesto, 2014–2019 for the new European Parliament and new commissioners. *Industrial Biotechnology*, 10(3), pp. 150–151.

EuropaBio, n.d.-a. GM benefits factsheet. In: EuropaBio, ed. n.d. Brussels: EuropaBio The European Association for Bioindustries.

EuropaBio, n.d.-b. Pocket guide to GM crops and policies. In: EuropaBio, ed. n.d. Brussels: EuropaBio The European Association for Bioindustries.

Express, T.I., 2013. Seeds of change. *The Indian Express*, 29 August. Available at: http://archive.indianexpress.com/news/seeds-of-change/1161406 [accessed 6 June 2016].

Frankman, E. and Weinberger, J., 2014. Vandana Shiva, voice of the anti-GMO debate. *The Takeaway*, 24 September. Available at: www.thetakeaway.org/story/vandana-shiva-voice-anti-gmo-debate/ [accessed 6 June 2016].

Gassmann, A.J., Petzold-Maxwell, J.L., Clifton, E.H., Dunbar, M.W., Hoffmann, A.M., Ingber, D.A. and Keweshan, R.S., 2014. Field-evolved resistance by western corn rootworm to multiple bacillus thuringiensies toxins in transgenic maize. *Proceedings of the National Academy of Sciences*, 8 April.

Gassmann, A.J., Petzold-Maxwell, J.L., Keweshan, R.S. and Dunbar, M.W., 2011. Field-evolved resistance to Bt maize by western corn rootworm. *Plos One*, 6(7). doi:e2262910.1371/journal.pone.0022629.

GeneCampaign, 2014. *GeneCampaign – about us*. [Online] available at: http://genecampaign.org/about-us [accessed 15 October 2014].

Gheysen, G., n.d. What is genetic modification? In: Bill Newton Dunn, ed. *booklet*. 2nd ed. Ghent: Bill Newton Dunn.

Giddens, A., 1999. Risk and responsibility. *The Modern Law Review*, 62(1), pp. 1–10.

Glover, D., 2008. Made by Monsanto: the corporate shaping of GM crops as a technology for the poor. *STEPS working paper*, 11.

Glover, D., 2010. Is Bt cotton a pro-poor technology? A review and critique of the empirical record. *Journal of Agrarian Change*, 10(4), pp. 482509. doi:10.1111/j.1471–0366. 2010.00283.x.

Greenpeace Nederland, 2012. *Roundup Readygewassen in Europa: Slecht nieuws voor boeren* [online] Amsterdam: Greenpeace Nederland. Available at: www.greenpeace.nl/ Global/nederland/2012/publicaties/GP_Benbrook_NL_def. [accessed 6 June 2016].

Grube, A., Donaldson, D., Kiely, T. and Wu, L., 2011. *Pesticide industry sales and usage 2006 and 2007 market estimates*. Washington, DC: US Environmental Protection Agency.

Hall, J.R., 1980. The time of history and the history of time. *History and Theory*, 19(2), pp. 113–131.

Hammond, E., 2010. *Counting the costs of genetic engineering*. Amsterdam: Greenpeace International.

Herring, R.J., 2006. Why did 'Operation Cremate Monsanto' fail? Science and class in India's great terminator-technology hoax. *Critical Asian Studies*, 38(4), pp. 467–493. doi:10.1080/14672710601073010.

ICRA, n.d. *ICRA: Institute for Cultural Research and Action*. [Online] www.hivos.nl/dut/ community/partner/30000114 [accessed 20 January 2015].

Jansen, K., 2004. Greening bananas and institutionalising environmentalism: self-regulation by fruit companies. In: K. Jansen and S. Vellema, eds. 2004. *Agribusiness and society. Corporate responses to environmentalism, market opportunities and public regulation*. London: Zed. pp. 145–175.

Jansen, K. and Vellema, S., 2004. Agribusiness and environmentalism: the politics of technology innovation and regulation. In: K. Jansen and S. Vellema, eds. 2004. *Agribusiness and society. Corporate responses to environmentalism, market opportunities and public regulation*. London: Zed. pp. 1–22.

Jasanoff, S. ed., 2004. *States of knowledge: the co-production of science and social order*. New York, NY: Routledge.

Jasanoff, S., 2011. *Frequently asked questions about sociotechnical imaginaries*. [Online] Available at: http://sts.hks.harvard.edu/research/platforms/imaginaries/imaginaries-faqs/ [accessed 27 October 2014].

Jasanoff, S. and Kim, S.H., 2009. Containing the atom: sociotechnical imaginaries and nuclear power in th United States and South Korea. *Minerva*, 47, pp. 119–146.

Jasanoff, S. and Kim, S.H., 2013. Sociotechnical imaginaries and national energy policies. *Science as Culture*, 22(2), pp. 189–196.

Jasanoff, S. and Kim, S.H., 2015. *Dreamscapes of modernity: sociotechnical imaginaries and the fabrication of power*. Chicago, IL, and London: University of Chicago Press.

Kathage, J. and Qaim, M., 2012. Economic impacts and impact dynamics of Bt (Bacillus thuringiensis) cotton in India. *Proceedings of the National Academy of Sciences of the United States of America*, 109(29), pp. 11652–11656. doi:10.1073/pnas.1203647109.

Krishna, V.V. and Qaim, M., 2012. Bt cotton and sustainability of pesticide reductions in India. *Agricultural Systems*, 107, pp. 47–55. doi:10.1016/j.agsy.2011.11.005.

Kuruganti, K., 2009. Bt cotton and the myth of enhanced yields. *Economic and Political Weekly*, XLIV(22), pp. 29–33.

Kuruganti, K., 2012. Interview, convenor Alliance for Sustainable and Holistic Agriculture (ASHA). Bangalore, 16 February 2012.

Kuruganti, K., 2013. Interview, convenor Alliance for a Sustainable and Holistic Agriculture (ASHA). Bangalore, 8 January 2013.

Kuruganti, K., 2015. *Food farmers freedom*. [Online] Available at: https://kavithakuruganti.wordpress.com [accessed 19 October 2015].

Levidow, L. and Papaioannou, T., 2013. State imaginaries of the public good: shaping UK innovation priorities for bioenergy. *Environmental Science & Policy*, 30, pp. 36–49.

Marcus, G.E. ed., 1995. *Technoscientific imaginaries: conversations, profiles, and memories*. Chicago, IL: Chicago University Press.

Monsanto CEO Grant discusses impact of genetically engineered crops on human health. 2008. [Online video] Producer: H. Grant. Available at: www.eenews.net/tv/videos/874 [accessed 28 August 2014].

Monsanto, 2012. Sustainability report. In: Monsanto Company, ed. 2012. St. Louis, Missouri: Monsanto Company.

Murugkar, M., Ramaswami, B. and Shelar, M., 2007. Competition and monopoly in Indian cotton seed market. *Economic and Political Weekly*, 42(37), pp. 3781–3789.

Padmanaban, G., 2012. Interview, Professor for Biotechnology, Indian Institute of Science, Bangalore, 19 February 2012.

Padmanaban, G., 2013. Sow the wind, reap a storm, opinion. *The Hindu*, [online] 2 September 2013. Available at: www.thehindu.com/opinion/op-ed/sow-the-wind-reap-a-storm/article5082915.ece [accessed 16 June 2016].

Paul, H., Steinbrecher, R., Michaels, L. and Kuyek, D., 2004. *Hungry corporations. Transnational biotech companies colonise the food chain*. London: Zed Books.

Pilson, D. and Prendeville, H.R., 2004. Ecological effects of transgenic crops and the escape of transgenes into wild populations. *Annual Review of Ecology, Evolution, and Systematics*, 35, pp. 149–174.

Pinch, T. and Bijker, W., 1987. The social construction of facts and artifacts: or how the sociology of science and the sociology of technology might benefit each other. In: W.E. Bijker, T.P. Hughes and T. Pinch, eds. 1987. *The social construction of technological systems. New direction in the sociology of technology*. Cambridge, MA: MIT Press. pp. 17–50.

Powerbase, 2010. Powerbase public interest investigations. *Profile: EuropaBio*. Available at: http://powerbase.info/index.php/EuropaBio [accessed 4 September 2014].

Powles, 2010. Gene amplification delivers glyphosate-resistant weed evolution. *PNAS*, 107(3), pp. 955–956.

Quartz, J., 2011. *Constructing agrarian alternatives. How a creative dissent project engages with the vlnerable livelihood conditions of marginal farmers in south India*. Maastricht: Universitaire Pers Maastricht.

Ramanjaneyulu, G.V., 2013. Interview, PhD Agricultural Extension, executive director, Centre for Sustainable Agriculture (CSA), Secunderabad, 19 January 2013.

Ranjith, M.T., Prabhuraj, A. and Srinivasa, Y.B., 2010. Survival and reproduction of natural populations of helicoverpa armigera on Bt-cotton hybrids in Raichur, India. *Current Science*, 99(11).

Rao, C.K., 2013a. Genetically engineered crops would ensure food security in India. In: D.J. Bennett and R.C. Jennings, eds. 2013. *Successful agricultural innovation in emerging economies. New genetic technologies for global food production*. Cambridge: Cambridge University Press.

Rao, C.K., 2013d. Interview, Professor Biotechnology, executive secretary, Foundation for Agricultural Biotechnology Awareness and Education (FABE), Bangalore, 10 January 2013.

Roseboro, K., 2013. The GMO seed monopoly: fewer choices, higher prices. *Food Democracy Now!*, 4 October. [Blog] available at: www.fooddemocracynow.org/blog/2013/oct/4/the_gmo_seed_monopoly_fewer_choices_higher_prices [accessed 20 October 2015].

Sahai, S., 2012. Interview, PhD Genomics, Founder Gene Campaign, New Delhi, 28 February 2012.

Saldanha, L., 2012. Interview, activist and lawyer, Environmental Support Group (ESG), Bangalore, 20 February 2012.

Schimpf, M., 2012. Interview, food campaigner, Friends of the Earth Europe, Brussels, 28 September 2012.

Schurman, R. and Munro, W.A., 2010. *Fighting for the future of food. Activists versus agribusiness in the struggle over biotechnology*. Minneapolis, MN: University of Minnesota Press.

Servaas du Marchie, C., 2012. Interview, director agricultural biotechnology division, European Association for Bioindustries (EuropaBio), Brussels, 10 December 2012.

Setharama, N., 2013. Interview, PhD Agricultural Plant Science, president, Association of Biotech-Led Enterprises (ABLE), New Delhi, 28 January 2013.

Sharma, S., 2013. Critical time. *Communication and Critical/Cultural Studies*, 10(2–3), pp. 312–318.

Shiva, V., 1988. *Staying alive: women, ecology and survival in India*. New Delhi: Kali for Women.

Shiva, V., 1991. *The violence of the green revolution*. London: Zed.

Shiva, V., 2000. *Stolen harvest: the hijacking of the global food supply*. Cambridge, MA: South End.

Shiva, V., 2014. Transcript, speech, Vandana Shiva at 'Food Otherwise Conference, 21 February 2014'. Available at: www.voedselanders.nl/voedselanders.nl/Conferentie_2014_files/FINAL%20transcript%20vandana%20shiva-%20plain%20text.pdf [accessed 21 August 2014].

Shiva, V., Barker, D. and Lockhart, C., 2011. *The GMO emperor has no clothes: a global citizen's report on the state of GMOs – False promises, failed technologies*. [Online] Florence: Navdanya, Centre for Food Safety and Navdanya International. Available at: www.navdanyainternational.it/images/pubblicazioni/The_GMO_Emperor_Has_No_Clothes.pdf [accessed 6 June 2016].

Singh, Y., 2013. Interview, leader, Bharatiya Kisan Union. New Delhi, 23 February 2013.

Smith, E., 2009. Imaginaries of development: the Rockefeller foundation and rice research. *Science as Culture*, 18(4), pp. 461–482.

Snow, A.A., Andow, D.A., Gepts, P., Hallerman, E.M., Power, A., Tiedje, J.M. and Wolfenbarger, L.L., 2005. Genetically engineered organisms and the environment: current status and recommendations. *Ecological Society of America*, 15(2), pp. 377–404.

Specter, M., 2014. Seeds of doubt. An activist's controversial crusade against genetically modified crops. *The New Yorker*, 25 August. Available at: www.newyorker.com/magazine/2014/08/25/seeds-of-doubt [Accessed 16 June 2016].

Steger, M.B. and James, P., 2013. Level of subjective globalization: Ideologies, imaginaries, ontologies. *Perspectives on Development and Technology*, 12, pp. 17–40.

Sud, S., 2013. The twist in the GM tale. *Business Standard*, 4 June. Available at: www.business-standard.com/article/opinion/the-twist-in-the-gm-tale-113060300906_1.html [accessed 16 June 2016].

The Economic Times, 2013. President asks ICAR to bring clarity of safety concerns of GM crops. *The Economic Times*, 16 July. Available at: http://articles.economictimes.indiatimes.com/2013-07-16/news/40613672_1_gm-crops-president-pranab-mukherjee-safety-concerns [accessed 16 June 2016].

van Bekkem, H., 2012. Interview, campaigner, Gentech and Duurzame Landbouw, Greenpeace Nederland, Amsterdam, 13 December 2012.

Visvanathan, S., 1997. Footnotes to Vavilov: an essay on gene diversity. *A carnival for science: essays on science, technology, and development*. Delhi, New York, NY: Oxford University Press. pp. 48–93.

Weinberg, A.M., 1966, 1991. Can technology replace social engineering? In: W.B. Thompson, ed. 1991. *Controlling technology: contemporary issues*. Buffalo, NY: Prometheus Books. pp. 41–48.

Zacune, J., 2011. *Who benefits from GM crops. An industry built on myths*. Amsterdam: Friends of the Earth International.

Zhao, J.H., Ho, P. and Azadi, H., 2010. Benefits of Bt cotton counterbalanced by secondary pests? Perceptions of ecological change in China. *Environmental Monitoring and Assessment*, 173(1–4), pp. 985–949.

6 Preserving landscapes and reordering science–society relations

Imagining the future in transdisciplinary sustainability research

Thomas Völker

Introduction

> How can we investigate provision for the vulnerability of society and nature? How can we take a responsible approach to tackling climate change and regional development? What chances are there for life if development progresses as it has done to date? What will the future costs of environmentally relevant action be and who will bear them? What uses, burdens and treatment can our space tolerate? What kind of science culture does a sustainable society need?

The quote above is taken from the website of the Austrian research funding scheme proVISION.[1] It introduces the mission statement of the programme and poses questions that are central to it. These questions tackle issues of development and its potential physical and social limits. They also address climate change and possibilities for responsible actions. Put simply, they aim at opening up a debate about which futures we deem desirable. At the very end of this quote, however, there is a nice little twist. The last in this series of questions indicates that a new *science culture* might be needed in order to answer all these questions and deal with contemporary challenges. Well-established ways of producing knowledge and technoscientific innovations – or what is often referred to as 'traditional' science – might not suffice for dealing with the challenges that lie ahead.

Bringing together environmental issues with debates about changing ways of knowledge production, the funding scheme proVISION places itself in a broader context of sustainability research policy in Austria that is rooted in the National Environmental Plan (NUP)[2] and anchored in the Austrian Strategy for Sustainable Development.[3] In doing so, the programme blends into an international trend of funding schemes and initiatives that aim at a 'democratization' of science (Nowotny, 2003) and the related idea that heterogeneous actors should be integrated into knowledge production practices. Such initiatives include but are not restricted to the td-net in Switzerland, MISTRA[4] in Sweden and the German programme for social-ecological research.[5]

These different programmes and initiatives share the idea that contemporary environmental challenges call for new ways of producing knowledge and for attempts to better integrate science and society. Additionally, they tend to stress the importance of knowledge production practices for actualizing desirable futures.

In this chapter I focus on such collectively held ideas or visions and on their role in reordering science–society relations by carving out a 'socio-scientific imaginary of preserving and preventing' that is co-constitutive with sustainability research in Austria. The funding scheme proVISION, with its emphasis on transdisciplinary research, is a perfect site to explore how certain imaginations of attainable futures are slowly assembled, rehearsed and stabilized and for studying how they become mutually constitutive with particular ideas about integrating science and society. In particular, the chapter addresses the following questions: how are desirable futures imagined in Austrian sustainability research? How do they relate to scientific (re-)orderings and how are new modes of knowledge production expected to contribute to their actualization?

Debating knowledge production and sustainability

ProVISION, as well as its predecessors, must be understood against the background of an ongoing debate taking place in both academia and policy about changing modes of producing knowledge. Scientific knowledge production in these debates is framed to different degrees as being deficient and no longer sufficient for dealing with contemporary societal challenges.

Three main strands of debate can be distinguished: to start with, there is the discussion that circles around the notion of 'Mode 2 knowledge production' (Gibbons, 1994; Nowotny, Scott and Gibbons, 2001), which basically makes the argument that scientific knowledge is increasingly produced according to extra-scientific rationales. Societal transformations, the authors argue in their diagnosis of contemporary science, lead to a growing 'contextualization' of science and to a transition from Mode 1 to Mode 2 science. Mode 2 science is supposed to 'open up' knowledge production as scientific knowledge; in addition to being valid and reliable, it also needs to be 'socially robust' (Nowotny, Scott and Gibbons, 2001). Therefore, heterogeneous actors enter processes of knowledge production, and the sites in which knowledge is produced multiply. Transdisciplinarity is described as one of the key characteristics of Mode 2 knowledge production (Nowotny, 2007).

Literature concerned with the concept of 'post-normal science' (Funtowicz and Ravetz, 1992, 1993), in contrast, focuses not so much on long-term changes in research environments, but – emanating from a critique of quantitative risk assessment – claims that there is a need for extended forms of knowledge production in situations where 'facts are uncertain, values in dispute, stakes high, and decisions urgent' (Funtowicz and Ravetz, 1993, p. 744). Funtowicz and Ravetz's point of departure is a particular kind of problem, and they ask for the knowledge we have at our disposal for finding the appropriate solution(s).

Stating that what they call a 'practice of normal, puzzle-solving science and technology' (Funtowicz and Ravetz, 1992, p. 268) cannot deal with these problems, Funtowicz and Ravetz call for a 'new political epistemology of science' (Funtowicz and Ravetz, 1992, p. 252). This new epistemology needs to be able to handle the uncertainties of scientific knowledge and allow for the acknowledgement of different perspectives and expertise. Activists, different sorts of publics, and practitioners need to be included in knowledge production, which needs to become able to integrate 'extended facts' (Funtowicz and Ravetz, 1993; Ravetz, 1999) such as anecdotes, not officially published information and the like in order to go beyond the distinction between value and fact.

A third strand of the debate focuses on the notion of a 'triple helix' (Etzkowitz and Leydesdorff, 1997) and directs attention to newly emerging relations between universities, industries and governments. The triple helix can be regarded as a heuristic or an analytical tool for looking at changing relations between these different actors on an institutional level. While this seems similar to the Mode 2 'diagnosis', Etzkowitz and Leydesdorff are not primarily interested in practices of producing knowledge but focus instead on the changing relations between different institutions and organizations in national innovation processes. This is why Hellström and Jacob describe the work in this line of thinking as 'focused on the organisational context of Mode 2' (Hellström and Jacob, 2000, p. 76).

The ideas expressed in these notions were taken up in science and research policy as visible in several national and international funding programmes that explicitly articulate the need for more collaborative ways of producing knowledge. Most recently, changing ways of producing and circulating knowledge are discussed in terms of 'responsible research and innovation' (Owen, Macnaghten and Stilgoe, 2012) and care logics in science (Felt, Barben *et al.*, 2013); they also inform attempts to situate 'science in society' or make science 'with' and 'for' society in European science and research policy.[6]

Questioning the practices of research and innovation and calling for a new 'science culture', such debates direct attention to potential benefits of public participation in decision-making. Such claims of opening up science to heterogeneous actors are often linked to concerns about contemporary environmental challenges and thus are often situated within the context of discussions about sustainability and sustainable development. This is hardly a surprise, considering this debate has, from the outset, been closely related to environmental problems such as the unintended side effects of technological innovation in our industrialized 'risk-society' (Beck, 1986; Adam, 1998). These notions are themselves heavily contested and have a broad variety of meanings (Hopwood, Mellor and O'Brien, 2005). While some argue that sustainability or sustainable development can provide a solution to contemporary problems, others are not so optimistic, arguing that the term 'sustainable development' is in itself a contradiction that stabilizes contemporary orderings and tends to obscure the role that technoscientific innovation itself plays in bringing about contemporary environmental problems (Robinson, 2004). Sustainability in this sense is itself a deeply political and normative concept that challenges modes and scales of governance (Dryzek *et al.*, 2003; Jasanoff and Martello, 2004; Dryzek, Norgaard and Schlosberg, 2013), as well as its institutional and organizational formations (Miller, 2001). Controversies about sustainability therefore direct attention to the contested nature of science, publics, power and nature and provide an intriguing site for studying the role of imaginaries in the co-production of science and society. As Robinson puts it, sustainability needs to be understood as 'the emergent property of a conversation about desired futures that is informed by some understanding of the ecological, social and economic consequences of different courses of action' (Robinson, 2004, p. 381).

This quote and its hint towards conversations and negotiations about desirable futures actually point at what I think is most interesting about these debates, which is that they indicate a changing imagination of science and its relation to society. This changing imagination is closely related to particular utopian as well as dystopian futures that are being collectively imagined. The debates thereby not only focus on the ways in which knowledge is produced for dealing with environmental challenges but simultaneously also address the actors that need to be involved and, in a broader sense, issues around the responsibility of science as well as of the legitimacy of scientific knowledge claims. Therefore, it is necessary to look at science policy as a site of governing innovation and of shaping the modes of knowledge production. This means that the governance of innovation entails not only a politics of what – the knowledge and knowledge-based technologies that are regarded as necessary – but also a politics of how – i.e. the very ways in which knowledge is produced. This of course has important implications for the everyday realities of scientists. This chapter therefore addresses the imaginative resources that are mobilized on a policy level and asks how futures are collectively imagined together with the ways of producing and circulating knowledge in the case of an Austrian research funding scheme.

Analysing collectively held imaginations: socio-scientific imaginaries

To explore the imaginative resources guiding transdisciplinary sustainability research in Austria, I will focus on collectively held imaginations about attainable futures and the ideas about the knowledge needed to actualize them. Imaginations, as I intend to use the term, do not refer to mere fantasies of single creative minds but to collective visions of futures. As such, they are embedded in concrete scientific and/or technological practices, while at the same time being constitutive for these practices.

Research on imagination in science and especially Sheila Jasanoff and Sang-Hyun Kim's notion of 'sociotechnical imaginaries' (Jasanoff and Kim, 2009, 2015) provide the conceptual background for this analysis. Imaginaries are defined as 'collectively held, institutionally stabilized, and publicly performed visions of desirable futures, animated by shared understandings of forms of social life and social order attainable through and supportive of advances in science and technology' (Jasanoff and Kim, 2015, p. 4).

In this quote, we see the crucial elements of imaginaries: they need to be *collectively held* and they tend to be more stable when they are *institutionalized* in some form. Moreover, imaginaries need to be *publicly performed* in order to become or remain stable. Such a focus on the need of collective imaginations to be performed directs attention to a dynamic understanding of imaginaries, their historical development and the work of assembling and rehearsing different imaginaries (Felt, 2015). Finally, imaginaries focus on *visions of desirable futures* that are entwined with ideas about social order and scientific and technological progress. To sum up, analysing imaginaries is about asking how particular epistemic, social and moral orderings are being stabilized through collectively shared ideas about the future. Understood in this way, imaginaries are a site for studying the co-production of science and society.

Imaginaries in that sense are visions of good and attainable futures that help to 'explore the sources of long-lasting cross-national variations in S&T policy' (Jasanoff and Kim, 2009, p. 120). Jasanoff and Kim use the term as a possibility of conceptually grasping the intertwinedness of imaginations about the future and national policymaking. Building on previous work concerned with imagining as a social practice and collective enterprise (see e.g. Fujimura, 2003; Taylor, 2004; Appadurai, 2006 [1990]), Jasanoff and Kim emphasize that 'the capacity to imagine futures is a crucial constitutive element in social and political life' (Jasanoff and Kim, 2009, p. 3). By conceptualizing them as instrumental and futuristic, Jasanoff and Kim frame imaginaries as a means of creating political and public will for following certain visions of desirable futures.

Although Jasanoff and Kim talk about both scientific and technological projects, their empirical work focuses very much on technoscience. As I showed above, recent debates about changing modes of knowledge production direct attention not only to the specific technologies and/or knowledge that is needed, but also stress the need to rethink the ways in which knowledge shall be produced and circulated, while additionally opening up the range of actors that are supposed to be involved in these practices. These debates therefore articulate a normative vision of scientific practice, exploring how science and society ought to be integrated.

This chapter argues that, in Austrian sustainability research, a 'socio-scientific imaginary of preserving and preventing' is currently in the making. This imaginary is positioned as an explicit alternative to a focus on technological fixes for sustainability issues. With this empirical reaccentuation, I intend to direct attention to how collectively held visions of particular attainable or dystopian futures of society are co-constitutive with ideas about the kinds of knowledge needed for dealing with the 'grand challenges of our time' (Lund Declaration), the practices that are suitable for its production and circulation, and the actors that are supposed to participate in these practices. It allows for focusing on (i) imaginations of desirable futures of society and at the same time (ii) the role of science, especially the alternative relationship of science and society, in actualizing such futures.

Through exploring these issues, this chapter addresses the desirable relations of science and society imagined in the context of transdisciplinary sustainability research and asks how science and society get co-produced through these imaginaries. Sustainability research in Austria and the funding scheme proVISION in particular offer a perfect opportunity to study how an imaginary is being assembled and stabilized against the background of a more dominant collective idea about how science and society ought to be integrated. It is therefore also a site to explore how imaginaries are contesting each other in manifold ways.

ProVISION – reimagining science–society relations

ProVISION[7] was a research funding programme of the former Austrian Federal Ministry of Science and Research.[8] The programme published two calls for projects, one in 2004 and one in 2007, had an overall budget of €7.6 million and funded a total of 36 projects.[9] At the time of writing this chapter, all proVISION-funded projects are finished and there will be no more calls for projects within the framework of this funding scheme. ProVISION was part of the Austrian 'Research for Sustainable Development' (FORNE) research strategy[10] and was therefore situated within the broader field of sustainability research and sustainable development. This becomes visible in the institutional ties of the funding scheme. One example for such institutional connections is proVISION's so-called 'partner programme', 'Technologies for Sustainable Development' (TSD),[11] run by the Federal Ministry of Transport, Innovation and Technology (BMVIT). This programme focuses on technological solutions for environmental challenges and is still ongoing, whereas proVISION was ended after publishing two calls for projects.

What makes this funding scheme interesting from an STS perspective is its explicit agenda to fund transdisciplinary projects in the area of sustainability research. In a nutshell, transdisciplinarity describes the normative demand to create spaces for collaboration between scientific and so-called 'extra-scientific' actors throughout different stages of research projects: from problem articulation and knowledge production to the dissemination and application of results. The funding scheme proVISION therefore provides us with an intriguing possibility of empirically exploring one attempt of putting into practice the call for new modes of knowledge production. At the same time, it is a fascinating site for studying an imaginary in the making – the attempt to establish and stabilize an alternative collective idea of integrating science and society. This attempt takes place against the background of a more dominant imaginary of science–society relations visible for example in proVISION's partner programme, TSD.

The analysis for this chapter builds on different proVISION programme documents. These include project calls, research principles as well as material from the website of the funding scheme. I will also draw on additional national and international policy documents that are mentioned within the proVISION documents and on documents produced within the proVISION-funded projects. Furthermore, I analysed a semi-structured interview with one of the programme managers and public statements of programme officials. Based on these different materials I intend to tell the story of the attempt to establish an alternative idea of science–society relations that becomes visible in different sites within Austrian sustainability research. The data was analysed using a grounded theory approach (Strauss and Corbin, 1998) and its basic coding principles. Additionally, I draw on more recent elaborations and reaccentuations of the approach that emphasize the importance of concrete situational contexts and the materialities of practices (Clarke, 2005; Charmaz, 2006). In particular, Clarke's situational analysis approach provides a sound methodological frame for exploring imaginations of attainable futures together with their social and institutional manifestations.

A socio-scientific imaginary of preserving and preventing

The imaginary I want to highlight is visible in the discourse of proVISION policy documents and related documents, but discursive spaces such as policy documents are of course not the only place to look for imaginaries. They can also be observed in material practices and in institutional arrangements such as funding schemes, the implementation and operation of which can be interpreted as 'active exercises of state power' (Jasanoff and Kim, 2009). This chapter therefore starts out with a focus on the institutional set-up of proVISION and the historical development of sustainability research in Austria. After that, it will explore in more detail the different aspects that constitute this particular socio-scientific imaginary: unprecedented problems calling for new ways of producing and circulating knowledge, different actors who are supposed to collaborate, ideas about a redistribution of responsibility and, finally, the more dystopian futures that are an integral part of this imaginary. In describing these different elements of the imaginary, the chapter will address a number of questions: how are the problems that call for a change in the culture of science and research being framed? Who is imagined to be potentially affected by these problems and who is supposed to be responsible for dealing with them? What subject positions and forms of collaboration are emerging together with certain visions of the future? Concluding this chapter, I will bring together the different aspects of this socio-scientific imaginary of preserving and preventing and argue that while this imaginary in the making can be understood as an attempt to develop an alternative vision of science–society relations, we need to pay careful attention to the subtle ways in which more 'traditional' boundaries are reintroduced and well-established conceptualizations of entities such as 'science' or 'society' are stabilized.

Historical development and institutional set-up

On a policy level,[12] the starting point for sustainability research in Austria was the establishment of the 'National Environmental Plan' (NUP).[13] This plan, which was implemented in 1994, was designed to establish the principle of sustainability in Austrian policy. The importance of continuous production and circulation of knowledge for solving environmental problems figures very prominently in this document. Additionally, the notion of a 'common responsibility' for our future is established in this document. Integration of science and society, however, is mainly understood in terms of raising 'awareness' within a seemingly homogenous 'public'.

One of the main consequences of the NUP was the introduction of the 'Research for Sustainable Development' research strategy (FORNE), which provides the framework for funding programmes in the area of sustainability research. The first programme of this kind was called 'Cultural Landscape Research' (CLR), which is often described by programme representatives as fundamentally important for the way proVISION was set up. And there are indeed clear overlaps concerning the collectively held imaginations about the relation of science and society and also concerning the actors involved in the funding scheme both on the side of policy and on the side of research. The Ministry of Science and Research depicts the programmes as closely related, seeing as they share the same *research principles* and because of how they differ from other research funding schemes:[14]

> Both programmes, KLF [this is the German acronym of CLR] and proVISION, are distinct from other thematic programmes due to the research principles: next to inter- and transdisciplinarity the programme requires internationality and anchorage in the regions. Clear, generally intelligible language is a further principle, which was already promoted by KLF.[15]

Two things are especially interesting in this quote: first, we can see a rehearsal process going on, which is visible both in the continuation of different funding schemes and in the fact that this relation is explicitly emphasized through shared principles. Second, the description of these funding schemes builds on an idea of similarity through difference: not only do these programmes share their principles; they are also similar in being *distinct* from other funding schemes. CLR and proVISION are both presented as alternative ways of doing research in comparison to other research programmes following a more traditional imaginary of science and research present mainly as something that CLR and proVISION do not want to do.

It is noteworthy, however, that there are also differences between the two funding schemes that illustrate a certain degree of development and therefore direct attention to the dynamic features of this imaginary. Collaboration in CLR was mainly framed in terms of interdisciplinarity, while proVISION centres more on transdisciplinarity and on the importance of *bridging* a gap between science and society.

Now, how was proVISION set up institutionally? Back in 2003, researchers and practitioners involved in sustainability research were invited to comment on a first outline of the funding scheme. This process is described as a 'participatory procedure' (PL01, p. 447) as statements on the funding scheme were requested from actors 'from Praxis and science' (PL01, p. 473). These statements were collected and used to refine the initial idea of the funding scheme as proposed by the programme coordinators. At the beginning of 2004, an official programme kick-off was organized for researchers and practitioners. In this way, a diverse set of actors was given the opportunity to participate in the setting up of the programme. ProVISION thus was framed from the very beginning as explicitly participatory and inclusive. This brief story about the procedure of setting up the programme directs attention to some of the core ideas about science–society integration guiding the funding scheme, such as the participation and the integration of actors. It is important to note here that these ideas build on the underlying assumption of a gap between science and the so-called 'social reality'[16] that is supposed to be 'bridged'.[17] This procedure is a performance of proVISION as a funding scheme that stands for an alternative way of doing science and research and an alternative way of integrating science and society.

The idea of being a different or alternative mode of producing and circulating knowledge that I described above is also visible in the way the funding scheme is positioned institutionally in relation to other funding schemes. ProVISION was a funding scheme of the former Austrian Ministry of Science and Research (BMWF). It is accompanied by yet clearly distinguished from its so-called 'partner programme', which is called Technologies for Sustainable Development (TSD). This programme is also devoted to sustainability research, but – being situated at the Federal Ministry of Transport, Innovation and Technology (BMVIT) – focuses mainly on technological developments. Not surprisingly, TSD builds on the idea of a technological fix, i.e. the development of technological solutions to contemporary environmental problems. ProVISION does not only follow a different approach; it is also positioned much more so as a necessary alternative to TSD's model of sustainability as it focuses on the production and circulation of knowledge for solving contemporary problems and actualizing desirable futures. Thus, there is a clear institutional boundary between sustainability as a technological issue and sustainability as a social issue. Thinking about Robinson's (2004) distinction between reformist and radical and conceptions of sustainability, one might assume that proVISION was introduced as a 'partner' to TSD in order to provide the latter aspect to TSD's work. However, this was not the case and proVISION also took a more reformist stance and did not call for fundamental changes.

The institutional situatedness of proVISION at the Ministry of Science and Research and the distinction between a technological and a knowledge-focused sustainability programme, however, does become consequential for the programme and the actors funded by it: proVISION consists of two different levels and respectively demands two different things simultaneously from researchers: first, it is about scientific projects, i.e. to produce and circulate (scientifically excellent) knowledge that can be used for sustainable development. To this end, extra-scientific actors are supposed to be integrated into the projects. Second, and this concerns mainly the scientists and researchers involved, it is about constantly reflecting on and thereby modifying the ways in which knowledge is produced. Thus, it is about collectively experimenting with a model of knowledge production that is expected to be sensitive to the different societal needs while at the same time aiming at satisfying contemporary criteria of scientific excellence.

Knowledge for solving the most urgent problems

The idea of promoting alternatives to current modes of producing knowledge is central to the funding scheme proVISION. This claim is closely related to the way in which problems are understood. Therefore, looking at how problems are framed within the proVISION programme documents provides a neat entry point for analysing this imaginary and its particular idea of science–society relations. When it comes to problems and challenges, proVISION can be understood as a rehearsal of previous narratives in sustainability research. Contemporary societies in this narration face new and unprecedented problems and challenges that cannot be solved by so-called 'traditional' science. What is especially interesting in the case of proVISION is the particular temporality of the problems:

> Focusing on sustainability, it is aimed at making knowledge available for solving the most urgent problems in provision for nature and society: adaptation to climate change and its consequences, suitable life and work models, responsible use of natural and industrial resources, and environmental protection.[18]

The quote above is taken from the programme website and briefly sketches the basic mission of the programme. Two things are striking here: first, the main goal of the programme is to *make knowledge available* for solving problems. Instead of merely focusing on knowledge production, the circulation of knowledge and the importance of knowledge availability for actualizing particular futures is also stressed. The socio-scientific imaginary guiding proVISION thus assumes or postulates a particular relation of knowledge, (political) action and attainable ways of living. Second, it is a particular kind of problem that needs to be solved, namely *urgent problems*. Urgency here is accompanied by the more anticipatory term *provision*, which is further specified through notions like *adaptation* and long-term *protection*. This focus on *urgency* creates a particular temporal relation between knowledge and action. It emphasizes the immediate present as the locus for actions.

In addition to being urgent and calling for immediate action, environmental challenges like '[c]limate change due to the greenhouse effect, shortage of water resources, relative reduction in agricultural area and global forest area ... soil degradation, species extinction, global development disparities and migration, urbanization and urban sprawl' are described as 'the emerging trends in this century'.[19] These phenomena are *emerging* as a distinct feature of *this century* and are thus depicted as novel and characteristic to our time. In this way, a clear temporal demarcation is introduced between contemporary scientists and previous generations of scientists, who were not forced to deal with these kinds of problems. The novelty of these problems partly lies in the scale of their possible effect. No matter where they originate, they can possibly become globally hazardous. ProVISION frames this as a 'global environmental crisis'.[20]

Through this framing of problems, a certain relation of globality and locality constitutes the proVISION narrative. Global problems have potentially disastrous effects for particular regions and thus call for solutions on a local level. This interrelation between global and local phenomena makes contemporary environmental challenges increasingly complex. This narration about the urgency, novelty and complexity of contemporary challenges portrays hitherto appropriate scientific practices as no longer sufficient for actualizing attainable futures and calls for new solutions.

What we see here can be described as the temporal and spatial constitution of environmental problems in proVISION and in sustainability contexts more generally. ProVISION projects mainly deal with problems that are framed as threats or as upcoming problems. This means that they are problems right now because they will be problems in the future. Prevention becomes a central mission of the programme: it is crucial to make the right decisions and take the necessary actions in the present in order to prevent latent dystopian futures from actualizing. Austria needs to be *protected* from these threats and the way to do so – in the proVISION imaginary – is to produce knowledge that enables *adaption to climate change*. Furthermore, these problems are unprecedented in their complexity and thus call for new approaches for solving them. This particular understanding of problems becomes constitutive for different collectives and thus for a reordering of science–society relations.

Reordering science–society relations

The way problems are described in proVISION leads to an epistemic reordering. In order to be able to make the right decisions and actions, a particular kind of knowledge is needed: anticipatory knowledge or knowledge about 'the future' that is supposed to enable particular actors to calculate possible effects of our actions in the present and thus take care of Austria's future *responsibly*. Scientific knowledge, however, is no longer regarded as a sufficient basis for dealing with contemporary problems. As a programme manager puts it, 'Knowledge from the life-world shall be melted with scientific knowledge'.[21] This idea clearly resonates with the more general imagination of alternative ways of producing knowledge, which pictures science as a 'part of public life' in the sense that science is supposed to 'play a significant role in decision processes'. In doing so, science is 'predestined to direct society's view to certain areas that it considers important'.[22]

By framing problems mainly as potential future threats and simultaneously reconfiguring epistemic orderings, certain collectives are stabilized and thus also a social reordering takes place:[23] a community of transdisciplinary sustainability researchers, who are supposed to deal with the problems depicted in the programme documents, is emerging together with a very particular version of Austria.

In proVISION, a particular imagination of Austria is established together with a certain relation of global problems and their local effects. It consists of threatening global developments that have not yet reached Austria. However, preventive actions need to be taken right now in order to preserve precious Austrian landscapes. To a large extent, the stabilization of this vision of Austria is achieved through pictures displayed on the programme's website and in the programme documents.[24] As Jasanoff (2001) points out, images are able to transcend cultural boundaries and can thus play an important role in stabilizing communities by acting as 'visual anchors'. Furthermore, Felt (2015, p. 117) reminds us that the use of pictures helps to nourish and stabilize particular imaginaries. A closer look at the visual language of proVISION therefore seems to be a promising path for exploring the imaginary at work here.

The pictures below are taken from the programme website and rehearse a very particular idea of Austria. They show Austrian landscapes that are not yet touched by problems such as climate change or scarcity of water resources. They are mostly not even touched by civilization, and, if civilization is visible, it is in the form of people enjoying nature or sustainable technologies such as windmills.

Figure 6.1 Pictures of Austrian landscapes from the proVISION website (© Peter Burgstaller).

What becomes visible in this selection of pictures is a vision of Austria completely free of the threats described in the proVISION documents. The interesting thing about this way of depicting Austria is the particular temporality that is constituted through these pictures. They simultaneously show an idealized present and what Jasanoff describes as 'desirable and desired futures' (Jasanoff, 2015, p. 4). The future of Austria depicted this way clearly follows a conservation ideal. This means that the present Austria as imagined by policymakers is green, healthy and free of problems. This imagination of the present, however, is at the same time an imagination of a desirable future. Austria is supposed to stay this way, or, as proVISION puts it, this Austria needs 'protection'[25] and 'responsible care'.[26] Therefore, these pictures also show Austria after the current threats have been dealt with. Yet, this green country is constantly endangered. Preservation becomes a core concern, considering that it is a fragile future that is depicted here. These pictures thus constitute a fundamentally different temporality compared to the programme documents of proVISION's 'partner programme', TSD, where the relation of present and future appears to be more stable, even to the point of disappearing altogether.

A second collective that is co-constitutive with the framing of problems in proVISION is the community of actors responsible for providing solutions and thus for actualizing attainable futures while preventing dystopian ones from becoming reality: transdisciplinary researchers and their partners from the so-called 'social reality'.[27] One of the central ideas of the programme is that knowledge needs to be produced by scientists and so-called 'non-scientific partners'[28] in a collaborative effort in order to become *available*. This idea is visible, for example, in the evaluation criteria for the second call of the programme. One of them is the number of so-called 'Praxispartners'[29] collaborating in the projects. The programme articulates guidelines concerning the overall number of Praxispartners that shall become part of the projects of the second call (BMWF, 2007, p. 10). The collective of the transdisciplinary researchers consists of researchers from different disciplines and Praxispartners, but who exactly are the Praxispartners and how are they imagined to contribute to the projects?

In brief, Praxispartners are actors from outside academia who are expected to hold a particular kind of knowledge that is distinct from scientific expertise. This kind of knowledge is often framed as experience or local expertise. Additionally, Praxispartners are ascribed a particular way of approaching problems. Besides adding different kinds of knowledge and expertise to the project teams, including Praxispartners supposedly also enables 'maximizing efficiency'.[30] This idea of efficiency can refer to pragmatically solving the problems at hand and is therefore expected to contribute to the overall goals of *adaption* and *protection* as depicted in the initial quote. Efficiency, however, can also concern the objective of *making knowledge available*. The underlying model of science–society relations interprets scientists as producers of knowledge and actors from social reality as responsible for matters of knowledge transfer. This model is visible in programme documents when transdisciplinarity is described as '[i]mplementing the results in practice' (BMBWK, 2005, p. 5), which is supposed to be 'part of the project work' (BMBWK, 2005, p. 5.).

In an interesting way, the socio-scientific imaginary guiding proVISION combines ideas about integration and separation of actors and collectives. Although the notion that heterogeneous actors are supposed to take part in knowledge production is rehearsed on several occasions, tasks and duties seem to be separated on different levels. In the following section, I will elaborate on this idea of integration and explore questions related to a redistribution of responsibility.

Redistributing responsibility

As I showed above, the need for new ways of producing knowledge is narratively constructed by referring to particular problems: problems that call for urgent solutions, are novel, more complex and threatening on a global scale. In order to deal with these 'grand challenges of our time',[31] knowledge needs to be produced differently. This claim also affects common ideas concerning the distribution of responsibility in transdisciplinary collaborations.

In proVISION, such a moral reordering is achieved through stories that mostly build on the premise of a separation of different areas and the depiction of particular relations and mobilities between them. To start with, science and society are portrayed as clearly separated entities or spheres. Science, however, is supposed to leave its 'ancestral place' (BMBWK, 2005, p. 5) to go to the places of 'societal practice' (BMBWK, 2005, p. 5) or into 'social reality' (BMBWK, 2005, p. 5.). These quotes refer to what can be called the spatial politics of proVISION. Such politics are present in at least two different ways: first, these narratives are about leaving scientific places such as universities and about going out to meet the partners in their own spaces. Thus, it addresses the spaces of encounter. Second, spatial politics refer to concrete research activities and outcomes, which need to be 'assigned to a physical place' (BMBWK, 2005, p. 12). Solutions shall be provided for particular regions and their concrete problems.

So what are the reasons for going out into *social reality*? What kind of knowledge can be produced in this way? The funding scheme's first call for projects states that transdisciplinary sustainability research shall 'have in view the vulnerability of society and nature in a precautionary way'; it shall 'influence climate change and spatial development responsibly' (BMBWK, 2004, p. 3). Developing precautionary outlooks is the task for science: science is supposed to elaborate methods for looking into the future, providing means for forecasting and knowledge about possible future effects of actions in the present. Again, we can see a temporality at play in this narrative, which simultaneously addresses both the current vulnerability of society and nature and possible future vulnerabilities. The notion *precaution* in the context of this funding scheme especially refers to the need to constantly account for possible threatening effects of our actions.

But ideas about precaution and responsibility in the documents also indicate that science is supposed to become active and that its task is not finished with producing knowledge. Far more science is supposed to take part in the circulation and application of the knowledge it produces. In this sense, this imaginary corresponds to other policy papers in the area of sustainability research, especially the Brundtland Report (World Commission on Environment and Development, 1987) and its famous take on responsibility, which is explicitly referenced in the FORNE strategy (Paula, Smoliner and Tiefenthaler, 2004): 'Humanity has the ability to make development sustainable to ensure that it meets the needs of the present without compromising the ability of future generations to meet their own needs' (World Commission on Environment and Development, 1987).

In this definition, the focus is of course on an ethically motivated temporal orientation towards the future. Additionally to the idea of responsibility represented in the quote above, proVISION refers to the idea of including extra-scientific partners into the process of producing knowledge. Responsibility can therefore be understood as 'socially robust' in the sense referred to by Nowotny (2003).

As I mentioned above, the relation of science and society that is imagined describes a moral obligation for science to become active and mobile. Scientists are supposed to leave *the ivory tower* and become *bridge builders*. I think it is important to note that concepts like these – notwithstanding their integrative concerns – subtly stabilize traditional power relations between scientific and extra-scientific actors in regard to their agency. The distinction between active scientists with a moral obligation to leave their universities and to build bridges, on the one hand, and, on the other, passive partners from a quite broadly defined 'social reality', who are potentially affected by contemporary environmental problems and in need of solutions, creates a very particular space of encounter and results in scientists and researchers being the dominant actors in the production and circulation of knowledge.

Questioning our **Zukunftsfähigkeit**

Debates concerning the framing of risks in policy documents concerned with sustainability in German-speaking contexts often circle around the notion of *Zukunftsfähigkeit*. The term is often used in sustainability policy in German-speaking countries. In the German version of the Agenda 21 policy document it is sometimes a translation of the term 'sustainability', for example in 'sustainable use of factors of production' (UNEP, 1992, p. 2.14.c). In contrast to the much more common translation *Nachhaltigkeit*, the term *Zukunftsfähigkeit* in German is interesting because it explicitly focuses on our ability or capability to build, create or actualize attainable futures, and thus opens up a range of intriguing questions: who or what can be future-capable and who can or even should make somebody or something *zukunftsfähig*? Are there particular practices that are future-capable? Is it something that can be learned or achieved? And, of course, the question that relates to the stakes and the way risks are framed: What happens if we fail to become *zukunftsfähig*?

A quote from the Austrian Strategy for Sustainable Development states:[32] 'Future-capability is becoming a central societal and political question in the face of global trends like climate change and the increasing consumption of resources and energy sources and the development of global population' (Paula, Smoliner and Tiefenthaler, 2004, p. 4).

In this quote, *Zukunftsfähigkeit* is described as the central question for society and politics. This means that societal and political actors should focus on anticipatory activities in the sense of an ethically motivated orientation towards the 'temporal realm beyond the senses' (Adam and Groves, 2007, p. 3). The FORNE strategy paper also talks about ensuring a 'liveable future' (Paula, Smoliner and Tiefenthaler, 2004, p. 3), which is characterized by a 'competitive science' (Paula, Smoliner and Tiefenthaler, 2004, p. 3), a 'dynamic business location Austria' (Paula, Smoliner and Tiefenthaler, 2004, p. 3), and a 'faultless environment' (Paula, Smoliner and Tiefenthaler, 2004, p. 3). Being *zukunftsfähig* refers not only to (global) environmental issues. In sustainability research narratives referred to by proVISION, the term relates to both economic success and scientific excellence. This is described as yielding a 'double dividend' (BMBWK, 2004), an idea that can already be found in proVISION's predecessor programme, CLR. *Zukunftsfähigkeit* in this sense frames sustainability research as a possibility to stabilize and strengthen Austria's position as both an innovation leader and as an economically prosperous country. This conceptualization of sustainable development is in line with a modest reformist line of thinking in the sustainability literature according to which minor adjustments to our current practices, regulations and institutional arrangements will suffice to actualize particular desirable futures (Robinson, 2004). Despite attempts to position proVISION as an alternative to other funding schemes in the area of sustainability research, its take on *Zukunftsfähigkeit* indicates a conservationist approach to dealing with environmental challenges.

Presenting *Zukunftsfähigkeit* as a question, however, can be read in a different, more pessimistic or dystopian way: what does it mean if somebody is not future-capable? One might speculate that, if humankind does not succeed in actualizing the future depicted in the programme documents, this may lead to the destruction of our living environment. This reading is supported by several passages in the funding scheme's second call for projects, which frames current actions in regard to their potential to threaten our livelihood in the long run. Thus, the problem is neither that the future is regarded as open in the sense of Adam and Groves (2007), nor simply that the present is expanding (Nowotny, 1995). What is at stake here is the very existence of our common future. In that sense, the notion can be read as being quite fatalistic. This also relates to a call for immediate action, because what is at stake here is the survival of mankind as a whole. As we can see, there is a complex interplay between more positive and very pessimistic or dystopian imaginations at play in narratives about *Zukunftsfähigkeit*. These imaginations not only call for immediate measures but also for very specific knowledge and ways of producing and circulating this knowledge. This focus on environmental threats relates to current debates about the notion of resilience that direct attention to the ability of particular systems to absorb and recover from shocks such as natural disasters (see e.g. Dryzek, Norgaard and Schlosberg, 2013). Similar to the particular understanding of sustainable development I described above, the concept of resilience in the context of proVISION, with its relation to concepts like adaption and mitigation, also tends to emphasize the preservation of the status quo of a given system; in so doing, it naturalizes environmental problems and their causes, which in turn bears the danger of prematurely closing down possibilities for debate (Evans, 2011).

Imaginings related to *Zukunftsfähigkeit* nicely show how risks are framed in two interrelated ways: first, becoming *zukunftsfähig* refers to a particular interconnectedness between science, innovation and economics. Sustainability research shall provide scientifically excellent results, which will lead to innovations that will strengthen Austria's economic position and help establish the country as an innovation leader. Potential conflicts between the concepts of sustainability and development in this economic sense are not addressed. Second, environmental problems like climate change or scarcity of resources threaten the livelihood on a global scale. In this sense, *Zukunftsfähigkeit* refers to the possibility that humankind is at risk of having no future at all.

Discussion: a socio-scientific imaginary of preserving and preventing

The aim of this contribution was to explore the emergence and stabilization of collectively held ideas of science–society relations in sustainability research in Austria. Drawing inspiration from Jasanoff and Kim's concept of 'sociotechnical imaginaries' (Jasanoff and Kim, 2009, 2015), I inquired into the imagined 'desirable futures' (Jasanoff, 2015, p. 4) or 'preferred ways of living' (Felt, 2015, p. 104) that are mutually constitutive with social and scientific orderings. This meant exploring how knowledge production and circulation has been imagined in Austrian sustainability research policy over the past decades and carving out a particular imaginary guiding this area of research policy: a socio-scientific imaginary of preserving and preventing. This imaginary depicts Austria as an extraordinarily 'green' and environmentally friendly place, a place that is imagined as being clearly distinct and even sealed off to some extent, while at the same time being threatened by the potential local effects of global environmental challenges. This imaginary maps out desirable futures together with the 'monsters' (Dennis, 2015) that need to be contained and emphasizes the need to preserve a present status in the face of potentially disastrous effects of global environmental problems. This is visible, for example, in ideas such as adaption or mitigation as future-oriented practices. It is noteworthy, however, that these are ideas mainly concerned with conservation. Even when change is promoted, as for example when it comes to an improvement of the quality of life or further economic growth, this is mostly thought of as the stabilization of a current process, the status quo with an added temporal dimension. The interesting thing, and the reason why I think of it as a socio-scientific imaginary, is that these imagined futures are closely entwined with a certain mode of producing and circulating knowledge. In order to keep Austria the way it is – extraordinarily 'green' and a leader of innovation in sustainability issues – a particular relation of science and society is needed for *making knowledge available* for political action. Knowledge is thus considered key for a liveable and prosperous future. In order to preserve and prevent, knowledge needs to be produced and circulated in new ways. Science needs to leave its infamous *ivory tower* and new relations between science and society are called for. Epistemic, social and moral reorderings therefore become co-constitutive with collectively held visions of attainable and dystopian futures. The imaginary in this sense relates to ideas about 'collective experimentation' (Felt, Igelsböck *et al.*, 2013, 2016) and 'real-world experiments' (Gross and Krohn, 2005; Gross, 2010), insofar as epistemic practices are open for experimentation to a set of heterogeneous actors and the consequences thereof directly affect the so-called 'real world' – particular communities and even entire Austrian regions. Through opening up spaces of knowledge production, notions of expertise (Collins and Evans, 2002; Jasanoff, 2003; Wynne, 2003) are also re-negotiated: Different actors are ascribed distinct forms of expertise that can become relevant for the production and circulation of knowledge.

These are the basic features of the imaginary that demarcate the funding programme from so-called 'traditional science'. They are manifest in the institutional set-up of the funding scheme proVISION and in the explicit attempt to establish participatory procedures. Simultaneously to promoting integration and participation, however, inner-scientific boundaries are drawn, according to which social sciences are responsible for producing transdisciplinary knowledge, whereas natural and engineering sciences are 'free' from such requirements. This practice of material and symbolic boundary-drawing stabilizes a difference between two imaginaries: a dominant and already-established imaginary of knowledge production and transfer and one promoting more 'democratic' (Nowotny, 2003) ways of producing knowledge that is performed as an alternative and is still in the process of rehearsal and stabilization, still in the making.

Throughout this chapter I traced the temporal development of this particular imaginary across different funding schemes and different instances of sustainability policy and showed how, for example, ideas about integration changed throughout its formation. This historical orientation directs attention to the fact that while imaginaries 'can originate in the visions of single individuals' (Jasanoff, 2015) they need to be collectively adopted in order to become consequential. To achieve this, they need to be publicly rehearsed by a broad variety of different actors and institutionally anchored in order to gain stability. Especially Felt has directed our attention to this kind of work and to the crucial role it plays in the stabilization of imaginaries. She emphasizes that 'it is important to trace the process of developing this imaginary across time' (Felt, 2015, p. 117). The socio-scientific imaginary I outlined in this chapter is itself not limited to the funding scheme proVISION but is closely tied to the broader area of sustainability research in Austria. This means that proVISION is only one moment in the historical process in which this imaginary is being assembled and stabilized. It had already gradually taken shape in proVISION's predecessor programme, CLR, and also in framework policies like FORNE and the National Environmental Plan, but this imaginary and its focus on collective experimentation is apparent not only in research policy; it can also be found in attempts of universities to reorder science–society relations, such as the so-called 'research platforms' of the University of Vienna[33] or the 'research campuses' in Germany.[34]

I want to close this contribution with a remark on potential future research. For tracing the (attempted) stabilization of this imaginary, I think it is important to consider not only historical or temporal but also its social dynamics: how do researchers and their Praxispartners translate the imaginative resources assembled in the proVISION documents and how do they appropriate these resources through blending them with other ideas about the integration of science and society they have at their disposal? Such an exploration of the stabilization of imaginaries needs to be especially sensitive to moments of contestation and together with that to the potential redrawing of traditional boundaries. Already in this brief description of the socio-scientific imaginary guiding sustainability research in Austria, various tensions between the desire to establish an alternative collective imagination of producing and circulating knowledge and more traditional demarcations between science and society became palpable, for example, in the idea of different kinds of knowledge and in the distinction between an active science and a more passive and homogeneous public or 'social reality'. An analysis of this kind would be especially suitable for asking how different imaginative resources become constitutive for particular research practices and how they are themselves reshaped in the process. Additionally, comparative studies of socio-scientific imaginaries and their relationship to sustainability research practices are needed. Although there are several similar funding schemes across Europe, especially in the German-speaking countries, studies about transdisciplinarity mostly focus on methodological (Pohl and Hirsch Hadorn, 2008; Nicolescu, 2010) and conceptual issues (Max-Neef, 2005). As a consequence, we are only beginning to understand more about the concrete practices of transdisciplinary research. Exploring the imaginative resources that are co-constitutive with these practices and asking how they figure into contemporary reorderings of science–society relations seem to be a promising path to follow.

Notes

1 Writing a book chapter, even if done by a single author, is always a collective endeavour to some extent. This holds true also for this contribution, which builds on data gathered in the context of the 'Transdisciplinarity as Culture and Practice' research project, carried out at the Department of Science and Technology Studies, University of Vienna, from 2009 to 2013. I want to thank my colleagues Ulrike Felt (PI), Judith Igelsböck and Andrea Schikowitz, who have, through numerous discussions and comments, contributed greatly to the ideas and arguments presented in this chapter. Furthermore, I want to acknowledge the contributions of the actors engaged in proVISION, who generously shared their experiences and time with us. Finally, thanks are also due to the reviewers as the final version of this contribution benefited enormously from their helpful and constructive comments and suggestions.

Project website: http://sts.univie.ac.at/en/research/completed-projects/transdisciplinarity-as-culture-and-practice [accessed 25 February 2015].

2 Austria's 'Nationaler Umweltplan'. Available at: www.cedar.at/data/nup/nup-english/index.html [accessed 30 April 2013].

3 'The Austrian Strategy for Sustainable Development'. Available at: www.nachhaltigkeit.at/assets/customer/Downloads/Strategie/strategie020709_en.pdf [accessed 30 September 2015].

4 Mistra website: www.mistra.org/en/mistra.html [accessed 30 September 2015].

5 FONA website: www.fona.de/de/19711 [accessed 30 September 2015].

6 Horizon 2020 website: http://ec.europa.eu/programmemes/horizon2020/en/h2020-section/science-and-society [accessed 28 October 2014].

7 ProVISION website: www.provision-research.at [accessed 14 February 2012].

8 The ministry has been restructured and is now called the Federal Ministry of Science, Research and Economy. Federal Ministry website: www.en.bmwfw.gv.at/Seiten/default.aspx [accessed 27 February 2015].

9 These numbers were presented at an event called 'Das 3hoch3 der Nachhaltigkeit' by a representative of the Austrian Federal Ministry of Science, Research and Economy in May 2011.

10 FORNE website: www.forne.at [accessed 27 February 2015].

11 TSD website: www.nachhaltigwirtschaften.at/english/index.html [accessed 27 February 2015].

12 There are relations between the Austrian anti-nuclear movement – described in a fascinating analysis by Felt (2015) – and sustainability research. While exploring these connections would be both interesting and worthwhile, it is not in the scope of this chapter.

13 Austria's 'Nationaler Umweltplan'. Available at: www.cedar.at/data/nup/nup-english/index.html [accessed 30 April 2013].

14 The quotes throughout this chapter are translations by the author.

15 Website of the Federal Ministry of Science and Research: www.bmwf.gv.at/startseite/forschung/national/programmeme_schwerpunkte/provision [accessed 8 April 2013].

16 ProVISION website, key issue 7. Available at: www.provision-research.at [accessed 14 February 2012].

17 ProVISION website: www.provision-research.at [accessed 14 February 2012].

18 ProVISION website: www.provision-research.at [accessed 8 August 2013].

19 ProVISION website: www.provision-research.at [accessed 8 August 2013].

20 ProVISION website: www.provision-research.at [accessed 8 August 2013].

21 ProVISION unterwegs. Available at: www.provision-research.at/proVISIONunterwegs/dl/ausstellung/080724_Programmemtrailer.mp4 [accessed 20 April 2013].

22 ProVISION website: www.provision-research.at [accessed 8 August 2013].

23 This argument was developed by my colleagues and myself in a similar fashion in Igelsböck, J., Felt, U., Schikowitz, A. and Völker, T., 2011. *Between entanglement and purification: participatory research imagined and practiced.* Presentation at the annual meeting of the Society for Social Studies of Science (4S), 2–5 November 2011, Cleveland, USA.

24 ProVISION website. Accessed 8 August 2013: www.provision-research.at.

25 ProVISION website: www.provision-research.at [accessed 8 August 2013].

26 ProVISION website: www.provision-research.at [accessed 8 August 2013].

27 ProVISION website: www.provision-research.at [accessed 8 August 2013].

28 ProVISION website: www.provision-research.at [accessed 8 August 2013].

29 The term 'Praxispartner' is used synonymously with the notion of 'extra-' or 'non-scientific partner'. It refers to actors that come from outside of science, the so-called 'praxis' (the distinction between practice and theory resonates in this term) and are supposed to collaborate as partners of the researchers in proVISION-funded projects.

30 ProVISION website: www.provision-research.at [accessed 8 August 2013].
31 Lund Declaration online source. Available at: http://era.gv.at/object/document/130/attach/1lund_declaration_final_version_9_july.pdf [accessed 3 October 2015].
32 FORNE website: www.forne.at [accessed 27 February 2015].
33 Information website of the University of Vienna concerning interdisciplinary research platforms. Available at: http://rektorat.univie.ac.at/en/research-platforms [accessed 19 August 2014].
34 Website of the RWTH Aachen: www.rwth-aachen.de/cms/root/Die-RWTH/Aktuell/Pressemitteilungen/September/~cwzc/Doppelter-Sieg-fuer-RWTH-bei-BMBF-Foerderi/?lidx=1 [accessed 19 August 2014].

References

Adam, B., 1998. *Timescapes of modernity: the environment and invisible hazards*. London and New York, NY: Routledge.

Adam, B. and Groves, C. eds., 2007. *Future matters. Action, knowledge, ethics* (Vol. 3). Leiden: Brill.

Appadurai, A., 2006 [1990]. Disjuncture and difference in the global cultural economy. In: M.G. Durham and D.M. Kellner, eds. 2007. *Media and cultural studies: key works*. Malden, Oxford and Victoria: Blackwell. pp. 584–603.

Beck, U., 1986. *Risikogesellschaft. Auf dem Weg in eine andere Moderne*. Frankfurt am Main: Suhrkamp.

BMBWK, 2004. *proVISION: Vorsorge für Natur und Gesellschaft. Erste Ausschreibung*. Vienna.

BMBWK, 2005. *proVISION: Vorsorge für Natur und Gesellschaft. Prinzipien*. Vienna.

BMWF, 2007. *proVISION: Vorsorge für Natur und Gesellschaft. Zweite Ausschreibung*. Vienna.

Charmaz, K., 2006. *Constructing grounded theory. A practical guide through qualitative analysis*. London/Thousand Oaks, CA/New Delhi: SAGE.

Clarke, A.E., 2005. *Situational analysis. Grounded theory after the postmodern turn*. Thousand Oaks, CA: SAGE.

Collins, H.M. and Evans, R., 2002. The third wave of science studies. *Social Studies of Science*, 32(2), pp. 235–296.

Dennis, M.A., 2015. Our monsters, our selves: reimagining the problem of knowledge in Cold War America. In: S. Jasanoff and S.-H. Kim, eds. 2015. *Dreamscapes of modernity: sociotechnical imaginaries and the fabrication of power*. Chicago, IL: Chicago University Press.

Dryzek, J.S., Downes, D., Hunold, C., Schlosberg, D. and Hernes, H.-K. eds., 2003. *Green states and social movements: environmentalism in the United States, United Kingdom, Germany, and Norway*. Oxford: Oxford University Press.

Dryzek, J.S., Norgaard, R.B. and Schlosberg, D., 2013. *Climate-challenged society*. Oxford and New York, NY: Oxford University Press.

Etzkowitz, H. and Leydesdorff, L., 1997. *Universities and the global knowledge economy: a triple helix of university – industry – government relations*. London: Printer.

Evans, J.P., 2011. Resilience, ecology and adaptation in the experimental city. *Transactions of the Institute of British Geographers*, 36(2), pp. 223–237.

Felt, U., 2015. Keeping technologies out: sociotechnical imaginaries and the formation of a national technopolitical identity. In: S. Jasanoff and S.-H. Kim, eds. 2015. *Dreamscapes of modernity: sociotechnical imaginaries and the fabrication of power*. Chicago, IL: Chicago University Press.

Felt, U., Barben, D., Irwin, A., Joly, P.-B., Rip, A., Stirling, A. and Stöckelová, T., 2013. *Science in society: caring for our futures in turbulent times*. Science Policy Briefing 50. Strasbourg: European Science Foundation.

Felt, U., Igelsböck, J., Schikowitz, A. and Völker, T., 2013. Growing into what? The (un-)disciplined socialisation of early stage researchers in transdisciplinary research. *Higher Education*, 65(4), pp. 511–524.

Felt, U., Igelsböck, J., Schikowitz, A. and Völker, T., 2016. Transdisciplinary sustainability research in practice: between imaginaries of collective experimentation and entrenched academic value orders. *Science, Technology & Human Values*. doi: 10.1177/0162243915626989.

Fujimura, J.H., 2003. Future imaginaries. Genome scientists as sociocultural entrepreneurs. In: A.H. Goodman, D. Heath and M.S. Lindee, eds. 2003. *Genetic nature/culture: anthropology and science beyond the culture divide*. Berkeley, CA: University of California Press.

Funtowicz, S. and Ravetz, J., 1992. *Three types of risk assessment and the emergence of post-normal science*. Westport, CT: Praeger.

Funtowicz, S. and Ravetz, J., 1993. Science for the post-normal age. *Futures*, 25(7), pp. 739–757.

Gibbons, M., 1994. The emergence of a new mode of knowledge production. In: *Workshop social studies of science in an international perspective*. Proceedings. Vienna, 13–14 January 1994. Vienna: University of Vienna.

Gross, M., 2010. *Ignorance and surprise: science, society, and ecological design*. Cambridge, MA, and London: MIT Press.

Gross, M. and Krohn, W., 2005. Society as experiment: sociological foundations for a self-experimental society. *History of the Human Sciences*, 18(2), pp. 63–86.

Hellström, T. and Jacob, M., 2000. Scientification of politic or politicization of science? Traditionalist science policy discourse and its quarrels with mode 2 epistemology. *Social Epistemology*, 14(1), pp. 69–77.

Hopwood, B., Mellor, M. and O'Brien, G., 2005. Sustainable development: mapping different approaches. *Sustainable Development*, 13(1), pp. 38–52.

Jasanoff, S., 2001. Image and imagination: the formation of global environmental consciousness. In: C.A. Miller and P.N. Edwards, eds. 2001. *Changing the atmosphere: expert knowledge and environmental governance*. Cambridge, MA, and London: MIT Press. pp. 309–337.

Jasanoff, S., 2003. Breaking the waves in science studies: comment on H.M. Collins and Robert Evans, 'The third wave of science studies'. *Social Studies of Science*, 33(3), pp. 389–400.

Jasanoff, S., 2015. Future imperfect: science, technology, and the imaginations of modernity. In: S. Jasanoff and S.-H. Kim, eds. 2015. *Dreamscapes of modernity: sociotechnical imaginaries and the fabrication of power*. Chicago, IL: Chicago University Press. pp. 1–47.

Jasanoff, S. and Kim, S.-H., 2009. Containing the atom: sociotechnical imaginaries and nuclear power in the U.S. and South Korea. *Minerva*, 47(2), pp. 119–146.

Jasanoff, S. and Kim, S.-H. eds., 2015. *Dreamscapes of modernity: sociotechnical imaginaries and the fabrication of power*. Chicago, IL: University of Chicago Press.

Jasanoff, S. and Martello, M.L., 2004. *Earthly politics: local and global in environmental governance*. Cambridge, MA, and London: MIT Press.

Max-Neef, M., 2005. Foundations of transdisciplinarity. *Ecological Economics*, 53(1), pp. 5–16.

Miller, C., 2001. Hybrid management: boundary organizations, science policy, and environmental governance in the climate regime. *Science, Technology & Human Values*, 26(4), pp. 478–500. doi:10.1177/016224390102600405.

Nicolescu, B., 2010. Methodology of transdisciplinarity–levels of reality, logic of the included middle and complexity. *Transdisciplinary Journal of Engineering & Science*, 1(1), pp. 18–37.

Nowotny, H., 1995. Wer bestimmt die Zeit? Zeitkonflikte in der technologischen Gesellschaft zwischen industrialisierter und individualisierter Zeit. In: K. Weis, ed. 1995. *Was ist Zeit? Zeit und Verantwortung in Wissenschaft, Technik und Religion*. Munich: dtv.

Nowotny, H., 2003. Democratising expertise and socially robust knowledge. *Science and Public Policy*, 30(3), pp. 151–156.

Nowotny, H., 2007. *The potential of transdisciplinarity*. [Online] available at: http://helga-nowotny.eu/downloads/helga_nowotny_b59.pdf [accessed 6 June 2016].

Nowotny, H., Scott, P. and Gibbons, M., 2001. *Re-thinking science. Knowledge and the public in an age of uncertainty*. Cambridge: Polity.

Owen, R., Macnaghten, P. and Stilgoe, J., 2012. Responsible research and innovation: from science in society to science for society, with society. *Science and Public Policy*, 39(6), pp. 751–760.

Paula, M., Smoliner, C. and Tiefenthaler, B., 2004. *Forschung für nachhaltige Entwicklung. FORNE. Rahmenstrategie 2004* plus. Vienna: BMBWK, BMVIT.

Pohl, C. and Hirsch Hadorn, G., 2008. Methodological challenges of transdisciplinary research. *Natures Sciences Sociétés*, 16, pp. 111–121.

Ravetz, J., 1999. What is post-normal science? *Futures*, 31, pp. 647–653.

Robinson, J., 2004. Squaring the circle? Some thoughts on the idea of sustainable development. *Ecological Economics*, 48(4), pp. 369–384.

Strauss, A.L. and Corbin, J.M., 1998. *Basics of qualitative research: techniques and procedures for developing grounded theory*. Thousand Oaks, CA: SAGE.

Taylor, C., 2004. *Modern social imaginaries*. Durham, NC: Duke University Press.

UNEP, 1992. *Rio Declaration on environment and development*. Available at: www.unep.org/documents.multilingual/default.asp?documentid=78&articleid=1163 [accessed 6 June 2016].

World Commission on Environment and Development, 1987. *Our common future*. Oxford: Oxford University Press.

Wynne, B., 2003. Seasick on the third wave? Subverting the hegemony of propositionalism: response to Collins & Evans (2002). *Social Studies of Science*, 33(3), pp. 401–417.

7 An automobile nation at the crossroads

Reimagining Germany's car society through the electrification of transportation[1]

Alexander Wentland

Introduction

After decades of relative stability within the motor industry and automobile culture, a considerable shift can be observed. The trend is unsteady but relates to fundamental and irreversible developments in the world. In the emerging economies such as China and India, highly populated areas are burgeoning into 'megacities', creating pollution and transportation issues of unprecedented proportions. Industrialized countries have been facing similar problems for decades, while their sporadic efforts to counteract the downsides of mass motorization mostly have come to nothing. By now, increasingly ambitious goals to mitigate climate change and the resulting pressure to modernize and decarbonize all areas of human activity have disrupted this inertia. The transition towards sustainable, decarbonized transportation systems has become an urgent matter around the world.

Policymakers in most countries consider the electrification of transportation with energy from renewable sources key to lowering emissions from transportation (IEA, 2009, 2014), but economic factors are deemed equally important. Leading car manufacturing countries such as the United States, Germany and Japan perceive this looming transformation as a threat to their national business models, but also as an opportunity to get an advantage in an anticipated future economy dominated by 'green' products and services. These countries have already experienced a few unsuccessful attempts to promote electric cars as low-emission alternatives to their conventional counterparts.[2] In contrast to the previous ill-fated efforts, this time, the technology, economic circumstances and consumer attitudes favour electrification. Even if it is too early to assume that a substantial share of vehicles will be electric anytime soon, the imagination around the 'car of the future' has never been more vivid.

Because of the perceived need to innovate, the image of the car and its design, purpose and meaning have become more contested than in previous times. The automobile's future has opened up and seems more uncertain than ever before. Some engineers, mobility experts and self-proclaimed futurists no longer envision a mere substitution of the internal combustion engine with an electric powertrain. They champion a radical departure towards novel scenarios, in particular the driverless car, as well as the seamless integration of electric vehicles (EVs) into electrical grids and communications networks (Wentland, 2016). It would not be the first time that changes in infrastructural arrangements and mobility practices would reinforce the course of economic and cultural development in a certain direction. This observation was as true for the formation of the suburban sprawl in the 1950s as it is for today's digitally mediated sharing economy. Both experiences reveal the density and contingency of the sociotechnical imagination around the automobile and how the seemingly mundane technical thing cannot be separated from political interests (Schwedes, Kettner and Tiedtke, 2013) and larger cultural dynamics (Ozaki, Shaw and Dodgson, 2013).

In fact, modernity itself has always been closely intertwined with the automobile. Contemporary social relations would not be possible without a highly sophisticated economy of car manufacturing, road infrastructure and traffic regulations. However, there is much more to it than that. The car has also become synonymous with what Charles Taylor (2004) identifies as the 'modern social imaginary', which emerges from the notions of progress, individualism and access to public life. Perhaps because society's basic fabric is so deeply imbued with the values, dreams and materiality of automobility, the story of the car exhibits few twists over the course of the twentieth century, despite a long-established awareness about its environmental impact. Now the story might be about to change.

In this chapter, I will explore recent changes in the electrification narrative, focusing on the particular case of Germany.[3] The country is and has always been an automobile nation in multiple ways. From the early history of mass motorization during the Third Reich to the global success of the German luxury car as a distinct brand and status symbol, the automotive industry has constituted the core of the nation's economy for almost a century. Simultaneously, to the German people, the automobile came to signify wealth, success, family values, pleasure, self-determination and a positive sense of nationhood as part of their collective post-Second World War identity. The conventional car has become part of Germany's national pride and central to its export-driven economic model.

Perhaps because Germany's industry, politics and culture are almost inseparable from the traditional regime of automobility, the electric car has been struggling to gain ground. Unlike in the United States, where within a decade Silicon Valley entrepreneur Elon Musk has built the multi-billion-dollar EV company Tesla Motors from scratch, or even compared to other European countries such as Norway or France, Germany's car manufacturers and consumers have remained sceptical. Against this backdrop, the German government passed one of the internationally most ambitious plans to electrify transportation. The 2009 National Initiative not only dedicated billions to research and development (R&D); it also brought together a heterogeneous alliance of stakeholders from business, politics and science to advance the joint project. Perhaps unintentionally, this questioning of the status quo paved the way for a broad range of imagined mobility futures associated with the EV into the discursive arena. Some of them represent a radical departure from the taken-for-granted notion of automobility.

These developments raise analytical questions with significant normative implications: what are the actor coalitions that co-produce and stabilize these imagined futures across different domains of society? What kind of desirable and obtainable future do they propose? How do they reimagine the car and its associated infrastructures as well as their users and mobility practices?

In this chapter, I argue that the electrification of transportation is more than an intended, manageable sustainability transition, as scholars in innovation studies usually analyse it (Geels, 2012). It also represents a rearticulation of the old promise of an automobile nation. Departing from this premise, I examine the recent attempt to re-envision automobility in Germany through the lens of 'sociotechnical imaginaries' (STI), a concept developed and elaborated by Sheila Jasanoff and Sang-Hyun Kim (2009, 2015). This kind of analysis seeks a more nuanced understanding of both technical transitions in the making as well as the institutionally and discursively sedimented habits of mind that highly technologized societies engender. After recapitulating some of the key features of the traditional German imaginary of automobility and its genealogy within its historical context, I will distinguish three major 'rearticulations' of future mobility. All of them draw together different assemblages of artefacts, infrastructures and users. They are being championed and stabilized by heterogeneous actor alliances. These coalitions partially overlap but are also competing to establish the dominant narrative in the larger discursive arena of the German energy transition.

Electrification, sociotechnical imagination and the nation state

Much of the recent literature has emphasized the economic momentum and institutional obstacles to a large-scale transition towards sustainable transportation but has only marginally taken into account the values, practices and desires attached to the automobile. Scholars in innovation studies have produced a number of insights on the prototyping and selection dynamics within the R&D of alternative fuels (Bakker, van Lente and Meeus, 2012). They also looked into economic prospects of users adopting EVs (Sierzchula *et al.*, 2014) and policy strategies such as government subsidies (Bakker and Farla, 2015). However, these studies seem limited in scope. They have avoided discussing the historical continuities and discontinuities, the nation-specific idiosyncrasies and the narratives and myths that allow engineers, policymakers and industry stakeholders to constitute and maintain a persistent 'system of automobility' (Urry, 2004).

Research from science and technology studies (STS) favour a different kind of perspective, paying closer attention to the interrelations between actor-networks, technical materialities and users. Unfortunately, little work on transportation has been done within STS.[4] The notable exceptions dealing with the electrification of transportation (Callon, 1980; Brown, 2001; Hård and Knie, 2001) remain analytically compelling but obviously cannot capture the far-reaching implications posed by the very recent developments in this area. In related disciplines, prominent works done by historians (Sachs, 1992; Volti, 2004; Seiler, 2008) and sociologists (Featherstone, 2004; Urry, 2007; Canzler, 2008) indicate a sustained interest in the history and culture of automobility in society as a whole. However, electric vehicles are usually excluded from the picture or discussed only as an early competitor for the combustion engine at the beginning of the twentieth century (Kirsch, 2000; Mom, 2004).

Other technologies have been studied in greater detail, leading to essential conceptual contributions. The history and sociology of technology have sensitized the social sciences to important aspects such as relevant social groups, interpretative flexibility and closure mechanisms, which enable and shape any given scientific and technological leap (Pinch and Bijker, 1984; Latour, 1986; Rammert, 1997). In particular, the work on large technical systems (Hughes, 1983; Mayntz and Hughes, 1988) and how such infrastructures reflect cultural preferences and historical shifts remains crucial to the understanding of current transitions towards sustainable technologies (van der Vleuten, 2004; Geels, 2007). However, this approach merits an extension. In order to probe the 'seamless web' (Hughes, 1986) that binds together the emerging EV technology and its infrastructures and society, the analysis has to aim attention at the desirable, collective futures that sustain such an interdependent network of materiality and meaning today.

The STI framework (Jasanoff and Kim, 2009, 2013, 2015) builds on the insights from the history and sociology of technology but adds the layer of collective, discursive and institutional co-production. It highlights the 'shared understandings of forms of social life and social order' (Jasanoff, 2015, p. 3) that technology is imbued with, but also stabilizes. This process can be observed particularly well on the level of the nation state. Recent empirical works have demonstrated how technologies contribute to the formation of national identities and how, vice versa, nation-specific cultures shape sociotechnical regimes (Ballo, 2015; Dennis, 2015; Felt, 2015; Felt, Schumann and Schwarz, 2015). These studies pay homage to Benedict Anderson's (1983/2006) influential definition of a nation as an 'imagined community', made possible by emerging technological opportunities. Anderson attributed the formation and stabilization of national imaginaries to the emergence of communication infrastructures all over the country, which allowed the people within a certain territory to develop a shared sense of belonging, even though they would never meet in person. He also reflected on the material manifestations of state power in and through sociotechnical devices of nation-making such as the map, census and museum.

Jasanoff and Kim have emphatically pointed to the state's role 'in defining the purposes of publicly supported science and technology', especially 'what constitutes the public good, which publics should be served by investments in S&T, who should participate in steering science and by what means, and how should controversies be resolved about the pace or direction of research and development' (Jasanoff and Kim, 2009, p. 120). The reciprocal problem of how 'national S&T projects encode and reinforce particular conceptions of what a nation stands for' deserves similar consideration. Both are two sides of the same coin, which renders the question of which came first meaningless. My analysis of the transportation transition follows the notion that national STI 'can penetrate the very designs and practices of scientific research and technological development' and that the resulting politics 'may shape not only the narrow issues surrounding those specific enterprises but also wider social and political understandings about a nation's past, present and future' (Jasanoff and Kim, 2009, p. 124).

Automobility represents an ideal case study for investigating deeper into a nation's STIs. This observation is particularly true these days, as the taken-for-granted regime around the internal combustion engine is being challenged by a variety of competing mobility futures based on the electrification of transportation. Each proposed mobility future reflects attitudes towards urban space, ownership and self-expression, as well as the technological preferences and materialities within a larger social landscape. Although the recent electrification efforts are to a large extent driven by global trends and transnational corporations, the STI perspective calls attention to the specificity of each nation's sociotechnical arrangements and experiences in the innovation process. The framework allows me to focus on the mutual stabilization of social imaginaries, technical artefacts and actor coalitions, within both Germany's historical entanglement with automobility and its current attempt to electrify transportation.

While the STI approach explicitly refrains from black-boxing 'innovation' or 'the state', this principle equally applies to the technopolitics of non-state actors such as corporations, NGOs and organized social groups. In the organizational field related to the competing technological rearticulations around the electrification of transportation, the state is only one actor among many. State politics are not necessarily coherent. In fact, the larger the actor, the more likely it is to occupy several – often contradictory – positions within a given field of imagination, which becomes especially apparent considering the position(s) that car manufacturers have taken. This finding is consistent with the assumption that a nation is not a monolithic entity but is composed of various cultural currents. Some of them are explicit and publicly articulated; some remain implicit or concealed. As a result, actor coalitions may align with different themes and currents within a national STI to advance their cause.

Although STIs appear to be rather stable pillars of the modern social order, the micropolitics, rehearsals and contestations should not be left out of the picture (Rammert, 2002, p. 183). Social imaginaries do not change abruptly, because they signify 'ways people imagine their social existence, how they fit together with others, how things go on between them and their fellows, the expectations that are normally met, and the deeper normative notions and images that underlie these expectations' (Taylor, 2004, p. 23). Technology introduces a disruptive force and dynamism into these seemingly firm collective dispositions. Scholars in STS have been particularly aware of this fact. Innovation can bring about rapid change and uncertainty that opens up seemingly taken-for-granted sociotechnical assemblages to the possibility of change. Emphasizing this point,

Gabrielle Hecht argues in her account of nuclear power in France that 'ideas about national identity do not grow by themselves', since they 'must be actively cultivated in order to persist'. As a consequence, 'articulating and rehearsing these ideas often reformulates them' (Hecht, 2009, p. 12). In the case of the competing rearticulations oscillating around the electrification of transportation, only their repeated mobilization and stabilization across various domains of society can prevent them from falling into oblivion (again). Likewise, the image of the car that many people hold dear had to evolve slowly over the course of Germany's troubled history in the twentieth century.

The promises of an automobile future

The ways in which modern Germany has imagined itself throughout its history are closely intertwined with the history of the automobile. Although the story of the car in Germany undoubtedly represents a success story, it also symbolizes the country's ambiguity towards what modernity stands for. At the turn of the twentieth century, the majority of people perceived cars as something foreign and strange, a passing trend in an unsettling time marked by rapid social and economic change (Volti, 2004, pp. 39–42). Kaiser Wilhelm II, who ruled Germany until the end of the First World War, notoriously stated, 'I believe in the horse, the automobile is merely a fad soon to disappear again'. Other contemporaries worried about motorization and, in particular, the escalating number of traffic fatalities. Many people equated the car with the violent surge of industrialization and urbanization, causing chaos and disturbance to their traditional lives. The anxiety did not last for long.

The emerging affection for the 'horseless carriage' was associated with two seemingly contrary predispositions, namely bourgeois romanticism and industrial tradition. Contemporary historians were able to identify the first traces of what later became the suburban lifestyle in the early romanticism around the car as a means of escapism to the countryside, which initially appeared as an activity in bourgeois circles in the first decades of the twentieth century (Sachs, 1992, p. 155). On the production side, early car manufacturing echoed Germany's economic tradition of industrial design, iron and steel craftsmanship, and manufacturing in general (Radkau, 1989, p. 86). Long before mass automobility, the STI around the car expressed key tensions within Germany's delayed path towards industrialized modernity (Kocka, 1988), tensions such as those between rural and urban life, tradition and progress, and economic inclusion and exclusion. It also gave a glimpse into the nexus between technology, industry and state politics that would emerge over the course of the century.

From the 1920s onwards, national policies pushed for mass motorization several times in German history. The proclaimed 'will to motorization' during the Third Reich gave birth to a comprehensive programme that encompassed the Fordist upscaling of car manufacturing, the nationwide development of highways and the promotion of cars as an affordable consumer product for every proud German citizen. The programme primarily aimed at boosting the economy and reducing unemployment. It also anticipated something that later became an integral part of the automobile society: the promise of equal access to the nation's wealth and opportunities. The iconic Volkswagen Beetle was the seductive expression for what came to be known as 'the people's car' (Rieger, 2013). The historian Wolfgang Sachs notes that,

> while in the twenties the automobile had conquered the desires of the masses, recreating a tense distance between those at the top and those at the bottom, the National Socialists now pledged to overcome the distance and help even the German worker get some wheels. [In doing so,] the regime lent recognition to the aspirations of the petty bourgeoisie in that it promised to strip the automobile of its luxury status in favour of a solid status as a use object.
>
> (Sachs, 1992, p. 56)

In retrospect, the National Socialists' imagined future of mass motorization can easily be dismissed as propaganda, but this would not do justice to the imaginative forces it unleashed. The programme had few consequences for actual sales figures in the 1930s. In fact, Germany fell further behind other countries in per capita motorization (Canzler, 1999, p. 28). Nevertheless, the state's effort commences the story of automobility as a kind of socioeconomic foundation for the nation. The VW Beetle and the desire for private transportation forged the modern STI of automobility and social mobility as two sides of the same coin. For the nation as a whole, the car became the manifestation of progress, technological achievement and territorial unification. The government motorization programme reinforced that sentiment, which lasted much longer than the Third Reich itself.

After the war, the newly born (West) German Federal Republic continued bolstering the car industry and encouraged private car ownership. Car manufacturers played a pivotal role in building the – partially mythical – narrative of the West German post-war 'economic miracle' (Radkau, 1989, p. 310). They created esteemed brands of cars for export and a multitude of jobs at home. The 'American Way of Life', imported during the time of the occupation and beyond, contained all that was needed to mould the desolate and disoriented country into a respectable, peaceful member of the then-emerging Western political and economic sphere. However, it took West Germany well into the 1950s before mass motorization became a real success within its own borders. The federal government subsidized domestic manufacturers while it also took every step possible to make car ownership more attractive, not only in terms of affordable prices.

The American model extended way beyond the restructuring of West Germany's industry and also inspired Germany's collective vision of a desirable and obtainable socioeconomic future (Burkart, 1994). The country's political parties unanimously agreed that the future belonged to the suburban 'home in the green'. This national consensus included social democrats, conservatives, liberals and even the Communist Party. When it came to the combination of home and car ownership, opposing ideologies could justify similar living arrangements while imagining fundamentally different forms of social life and social order. The ideal of the serene home outside the city supported the conservative notion of economic, political and cultural integration into the middle class. At the same time, home and car ownership catered to the idea of autonomy and self-determination using private property, which was explicitly embraced as 'the cultural will' of the working class by the Social Democratic Party (Polster and Voy, 1991, p. 303). From the moment that the state established this political consensus, it did not take long before the STI of automobility spread into other social realms.

Among urban planners and architects, the paradigm of the 'automotive city' mirrored the economic developments. As in many other Western countries, modernist philosophies of urban infrastructure gained momentum. In West Germany, one of its key figures was Hans-Bernhard Reichow, who coined the influential term 'auto-friendly city'. Similar to the ideas of the Swiss-French designer and urban planner Le Corbusier, Reichow's (1959) vision conceived the modern city as a complex organism, echoing the penchant for societal holism and technocratic functionalism during that period. Reichow perceived the traditional European city as disharmonious and hampered by obsolete material arrangements, which were no longer suitable for the era of mass automobility. Those outdated settlement structures, with cluttered housing and shabby streets, had to make room for a scientifically organized city, which would drastically improve life standards by separating work, living and consumption. At first, urban planners started realizing this vision in infrastructure projects around the country. Contrary to the United States, where this model prevailed quickly after the war and materialized in the urban sprawl, in West Germany the implementation of the auto-friendly city stopped halfway. The country settled for a much more moderate version of this modernist scheme. The majority of the population hung on to the traditional structures of the old city centres but embraced the promises attached to the functionalist penchant for private house and car ownership (Polster and Voy, 1991, pp. 309–310).

The system of automobility resulting from these political and cultural dynamics was also fuelled by a political economy of desires, where a tangible consumer product could signify societal aspirations and late industrial romanticism. Between the early 1960s and the first oil crisis in 1973, the number of registered cars quadrupled, the number of driven kilometres tripled and the length of the German autobahn system doubled. The government actively facilitated this process. It cut taxes on car ownership, subsidized home building and invested heavily in infrastructural development. A powerful STI of what the automobile meant for the nation engendered the state's involvement with mass motorization. For the recuperating West German society and its people, owning a car was synonymous with territorial unity (through the sophisticated highway system), societal access (through increased spatial mobility), economic self-esteem (through conspicuous consumption) and national pride (through the identification with West Germany as a global economic force).

Auto manufacturers and the growing car lobby jointly provided the language and imagery for this symbolic citizenship of automobility in the German version of what the cultural historian Cotton Seiler (2008) called the American 'republic of drivers'. Since its foundation at the beginning of the century, the General German Automobile Club (ADAC), which counted 18 million members in 2012 out of a national population of 81 million people, has launched several influential ad campaigns and even mobilized protests against government regulations. In their 1965 'Manifesto on Motorized Travel', the ADAC proclaimed automobility to be a universal right, in its words an 'indispensable means for satisfying basic needs' in a 'free and progressive society'. For that reason, there is 'an obligation to solve the great problems ahead of us as a nation, which stems from the dynamic between technological progress and the economic prosperity that results from it'. This obligation concerns the state in particular. According to the organization, the government is constitutionally as well as morally bound to pursue a progressive, car-friendly society in order to achieve national welfare but also to 'nourish and deepen the friendship between the peoples'[5] (ADAC, 1965, translated from Sachs, 1984, pp. 94–97).

Even though the traditional STI of automobility began to wane in the 1970s, it remains powerful. Traffic has lost its environmental innocence. Since the dramatic rise in air pollution and the two oil crises, it has become an administrative and political issue (Canzler, 1999, p. 19). The green movement and a new generation of urban planners have challenged the German car industry at home, as have new competitors abroad. Nevertheless, automobility still sustains key discourses of individual well-being and national identity. It has managed to integrate the conflicting poles of Germany's (now reunified) modern sociotechnical narrative, including the tensions between socioeconomic classes, between urban and rural living, individualism and belonging, and technological progress and historical continuity. The modern social imaginary, as Taylor describes it, powered by its key notions of national progress, individualism and equal access, has entrenched the car within society (Rammler, 2008). The car has come to represent all these promises and has given them a tangible, sociotechnical reality, even as concerns about its sustainability were widely acknowledged on a more reflexive level of social discourse (Beckmann, 2001). This strong cultural bond has made it difficult to modernize private transportation and is still considered to be the biggest obstacle to the electrification of transportation today.

Emerging crisis and possible departure from the previous path

In 2015, the Volkswagen emissions scandal was a symbolic earthquake for the combustion engine-based auto industry. The secret manipulation of the software in several diesel engine-based models, which made the vehicles appear significantly cleaner during emissions testing compared to their real benchmarks, shattered consumer trust in the brand. It also reassured the proponents of alternative fuels that conventional engines cannot keep up with tightening regulations. The scandal resulted in a legal battle in the USA and other countries, putting additional pressure on Volkswagen and its competitors. Meanwhile, German regulators and the general public remained relatively calm. The diverging perceptions of the deliberate manipulations have also shown how intimate the relationship between the German business interests, engineering cultures and national sentiments still is. While state officials and intergovernmental organizations across the industrialized world agree on the inevitability of substantial change, Germany as a country seems somewhat reluctant to follow through.

The Volkswagen scandal is only the most recent event in a jolting but ongoing shift in the STI of automobility. Defending the traditional regime has become increasingly difficult over the past decades. Although no sea change can be discerned yet and car manufacturers have continued to succeed with their fossil fuel-based models, German environmentalists and transportation experts have been advocating for an electrified transport network based on renewable energy sources. The transition towards a more sustainable, decarbonized society has become an urgent matter around the world as well. Car makers had previously pacified environmental concerns through incremental technical innovations such as the catalytic converter, efficiency gains in the internal combustion engine and more sophisticated traffic regulations. However, this strategy has started to show its costs and limitations. In the mid-2000s, the European Union committed the German government to ambitious greenhouse gas reduction plans. The consistently high oil price at that time put additional strain on the economy. Most importantly, the government was afraid that German car manufacturers could lose their edge on the global market, especially in the emerging Asian economies, if they did not invest in the electrification of their next-generation models (NPE, 2010b, p. 9).

Even though there is no consensus on how to achieve a carbon-free transportation system, the vast majority of experts and intergovernmental institutions agree that an immediate electrification of transportation must play a crucial role in this transition (European Commission, 2008; IEA, 2009, 2012). This assessment puts Germany in a difficult position. The leading car-producing nation, particularly esteemed for its fuel-consuming premium car brands, perceives any alteration of its key industry's business model as costly and risky. The German government has found itself in a paradoxical role in which it is urged to stimulate technological innovation while still undeniably being captivated by the traditional system of automobility, manufacturing and infrastructure.

Despite some ill-fated attempts to inject momentum into EV development in the 1980s and 1990s (Hård and Knie, 2001), the issue of electrification has once again become a central concern in Germany's industrial innovation policy. In 2009, the National Initiative, composed of representatives from industry, science and the state, set out to turn Germany into an EV 'lead market', bringing one million plug-in electric cars to the streets by 2020. The country also aimed at becoming the 'lead producer' for battery cars on the global market, investing 1.5 billion euros towards that end (Bundesregierung, 2009, 2011). Unlike in previous episodes, the outlook for a breakthrough on the market looks somewhat brighter now: the technical components, especially the battery, have become more sophisticated, cost-effective and versatile (Catenacci *et al.*, 2013). Cities and states around the world have introduced traffic privileges, tax discounts and other economic incentives for these vehicles. Even consumer attitudes have slowly come to view EVs in a positive light. In Germany, however, the shift to what has been dubbed 'electric mobility' (*Elektromobilität*) appears to be a laggard compared to the USA, China, Japan, France and other countries in absolute sales numbers as well as marketing by the countries' respective manufacturers.[6]

The breakthrough on the domestic market, in particular, seems to be a long time coming. In 2014, the market share remained at a meagre 0.4 per cent, with an estimated 24,000 EVs on the roads (NPE, 2014, p. 4), which is a low absolute number, even compared to much smaller countries such as Norway or Denmark. As an accepted and widely diffused innovation (Rogers, 1962/1983), the electrification of transportation in Germany has not yet reached its tipping point. German customers seem to be sceptical or ignorant of the fully battery-powered models that are currently available. They continue to regard EVs as far too expensive and limited in their capabilities (Franke and Krems, 2013). As a consequence, when talking about the progress made towards sustainable private transport, most media outlets frame it as both a failure of German car makers to deliver convincing products and a crisis of Germany's industrial and transportation policy. Experts criticize the National Platform, the formerly praised coalition of key stakeholders from industry, research and government, for being little more than a paper tiger. In their view, the corporatist institutions set up in the late 2000s mostly legitimized industry and research subsidies but did not offer a vision for a major transformation of the current system of automobility. Radical alternatives have been articulated by different actors, often with a much longer time frame in mind.

Even though presently it remains uncertain when electric cars will turn a profit, the imagination around the 'car of the future' has become more vibrant and more contested, especially within certain parts of the German expert community. With converging technologies and new corporate alliances, the design, purpose and meaning of the automobile seem to be open to reinterpretation. This ambiguity is partially due to the elusiveness of the object itself. Within the current discourse, one seemingly simple question is rarely asked: what does the 'electrification of transportation' mean? Such a definition would provide a frame of reference for the imagination of expert communities and potential users alike while highlighting certain technologies and application scenarios that deserve further attention and investment. The electric powertrain allows for different systems of energy storage, which can be a battery, a fuel cell, hydrogen or any fuel transformable into electricity during driving. In fact, the storage component is the most elaborate, dangerous, expensive and profitable part of an EV. All solutions require different infrastructures for refuelling or recharging. Which one will prevail?

While some actors deliberately aim at a 'technological fix' (Rosner, 2004) – that is, a substitution of some elements of the car – others try to break away from the established pathway of automobility by broadening the technological vision. They strategically rearticulate key themes of the STI around the car, evoking new images and desires for a new 'republic of drivers'. Automobility constitutes the perfect arena for futuristic reimaginations, as it has always represented much more than just a sophisticated machine or consumer product. It is a collective and distributed process, contingent upon intricate arrangements of persons, things, and signs (Rammert, 2012). Nonetheless, after so many decades, automobility has become a taken-for-granted part of modern life, a black box that is now being reopened through a more serious attempt at electrifying transportation. The EV entails an elaborate sociomaterial ecosystem of charging and maintenance, not to mention the need for new professional groups, business models and configurations of routines in everyday life (Shove, Pantzar and Watson, 2012). This situation creates a moment of possibility within the imagination around the car not only with regard to the decision between competing drive technologies but to different assemblages of actors, artefacts and practices (Oudshoorn and Pinch, 2003).

Towards a new STI of automobility: three competing 'rearticulations'

Although the old economic regime and cultural image of the conventional car persist in the media as well as on the streets, an array of competing mobility futures has emerged. Because the 2009 National Initiative could not keep up with expectations, other players have been able to fill the resulting vacuum of imagination with more radical ideas. Car manufacturers have been unwilling to move away from business as usual unless compelled to do so. The National Platform, the governance unit for the original stakeholder alliance, crumbled in the wake of disappointing sales numbers. This erosion gave way to the formation of new actor coalitions, including companies and organizations that had no previous association with transportation. Electrical grid operators, utilities and information and communication (ICT) companies gradually entered the stage with a different set of goals and expectations. NGOs, mobility experts and dedicated 'visioneers' (McCray, 2012) have been paving the way for them already by championing alternative transportation scenarios for decades. Many of the new corporate players have shown little interest in participating in the current EV market, but instead have rearticulated the STI of automobility for the next generation's technology. Meanwhile, this volatile situation allowed the car industry to adapt to the challengers, partially reforming aspects of the existing regime while drawing on traditional stories and values attached to the automobile.

The brief history of automobility in Germany provided above demonstrated that the STI of automobility has played a significant role in shaping current attitudes towards urban space, ownership and self-expression. In turn, social practices across all areas of life have nourished the collective sociotechnical imagination around the car for almost a century. This mutual interdependence has produced a very stable set of technological preferences – for engineers, policymakers and users alike – and created a seemingly inescapable situation within Germany's larger economic and cultural landscape. The proposed electrification of transportation should be viewed against this backdrop. The STI of automobility is ingrained in the forms of social life and social order in contemporary German society. Reassembling automobility is a driving force in social change, and, at the same time, social change is a condition for a successful technological transition.

I identify three major rearticulations of mobility, by which I mean three distinct visions of what a desirable system of (electrified) automobility should look like, as well as how and to what end the current structures would be reformed.[7] All of these rearticulations shape different assemblages of artefacts, infrastructures and users by creating different kinds of relationships between them. These emerging sociotechnical configurations are being championed and stabilized by heterogeneous alliances of actors. As a result, they are distinct in scale and force, especially in light of the actors' sources of capital, the degree of institutional support and how they currently enact their scenarios. Each evaluated rearticulation has to align with certain aspects of the traditional image of automobility as well, which is full of tension and ambiguity in itself.

The three rearticulations of the STI of automobility promote alternative sociotechnical assemblages, some aligning with the existing system of automobility and some going beyond the traditional understanding of the car, addressing fundamental shifts within contemporary societies. The first rearticulation can be summarized as 'fixing the technology' (Rosner, 2004), which includes a set of technologies and actors that is in many ways heterogeneous but also unified by the belief in some invention, ideally a technological breakthrough. For reasons I will further elaborate on, this can be seen as the most conservative articulation of future mobility or, in fact, a continuation of the previous system and its promises with some adjustments. The other two are more radical and broader in scope. The second rearticulation assumes that the energy transition and the electrification of transportation constitute natural allies. The actors articulating this vision try to 'find a new purpose' for the EV. They propose to turn the battery car into a functional part of the emerging 'smart grid' infrastructure. The third rearticulation moves the focus away from the technology to the conscious citizen. In this imagined future, drivers break away from the paradigm of combined car and home ownership by joining the post-materialist 'sharing economy' and using alternative means of transportation.

Rearticulating the old promise: fixing the technology

All previous attempts to position the EV as a viable alternative to the conventional car were not only doomed to fail; they also had one thing in common: they assumed that consumers had a specific image of a desirable car in mind, an image they were used to and, thus, not willing to change. As a consequence, most of the engineering around the EV tried to emulate the capabilities of conventional vehicles. In many cases, it continues to do so. The majority of current

EVs are refurbished versions of popular models, in which engineers have replaced the internal combustion engine and fuel tank with an electric powertrain and battery. Ideally, within the paradigm of so-called 'conversion design', both the manufacturer and the driver would have to adapt as little as possible. It is therefore not surprising that, compared to the range and refuel speed of a conventional car, no EV could ever stand a chance.

General Motors challenged this paradigm for the first time with its 'EV1', the first car systematically designed as an electric vehicle, emphasizing the light coachwork, unique driving features and futuristic design elements. This design was celebrated as exceptional industrial R&D and, at least from the engineer's viewpoint, as a liberation from the burden of history: The electric car of the future did not have to be identical to the internal combustion engine-based car of the past anymore. The STI of automobility, nonetheless, was still limited to individualized modes of transportation suitable for modern family life within Californian suburbia. Today, the EV1 has almost been forgotten due to its short commercial life. Engineers and industry experts rarely refer to GM's pioneering vehicle as a point of departure. However, the idea of purpose design, which acknowledges the fact that EV technology has specific strengths and weaknesses, has become the dominant paradigm in the engineering process.

In the last couple of years, car manufacturers have started introducing distinct models that they specifically designed to drive electric. Being aware of the fact that they are marketing expensive and, in the opinion of many consumers, inferior products, most manufacturers are trying to convey an 'aura of high tech' (Henderson, 1998). By using touchscreens, advanced driving assistance systems and recognizable designs, they encourage conspicuous consumption among environmentally concerned drivers and simultaneously try to enrol technophile early adopters in an exciting new trend. Tesla Motor's high-tech luxury cars, the 2012 'Model S' and 2015 'Model X', are prime examples of this commercial strategy. Some German automakers, especially BMW and Mercedes, have quickly adopted this scheme, while the Silicon Valley start-up is progressing to the next stage of its expansion plan. Tesla announced that it would begin production and delivery of its mass market-oriented 'Model 3' in 2017, promising the range and features of its premium counterparts but at a much lower retail price. Tesla's competitors are yet again under pressure to catch up by delivering cars that can also convince more traditional customers.

The green high-tech luxury approach has been internationally successful within a small niche market, but, for the envisioned large-scale electrification, most industry experts agree that there has to be rapid technological progress. In particular, the limited capacity of battery technology is perceived as a bottleneck. Electrical engineers argue that with the current lithium-ion technology, only incremental improvements can be achieved towards cheaper and more powerful batteries. Ideally, a breakthrough innovation could bring an electrochemical cell type into play that would address all the crucial issues impeding progress towards electrification, namely range, costs and safety. The myth of the possibility of a 'miracle battery' dates back to the early twentieth century, when

Thomas Edison wanted to outrace petroleum cars, which were, at the time, perceived as more forceful and long-lasting 'adventure machines' (Mom, 1999, p. 19). Needless to say, Edison, although making significant inventions along the way, could not 'fix the technology' to the extent he intended, measured by the standard set by the combustion engine. The petroleum car triumphed, while the battery vehicle vanished.

Later R&D projects, such as the German Rugia experiment (1992–1996), concluded that battery technology was still not yet ready to take over the streets. Echoing this belief, in its National Initiative the German government put the Fraunhofer Society, one of Europe's largest application-oriented research networks, in charge of pushing electrochemical research and exploring the effects of scale for domestic mass production. It also created a distinct task force on battery R&D within the National Platform. When the National Initiative started in 2009, many experts were hopeful they would see battery technology fixed sooner or later. They assumed the boom of mobile electronics would be a catalyser for advanced R&D in electrochemistry, a research field that has been, according to the common narrative, stagnating for many decades, at least in Germany. While being conservative in its general predictions, the EV battery task force deliberately maintained the hope that a post-lithium-ion technology was feasible within a decade and could vastly exceed the limitations of today's batteries (NPE, 2010a). The German government was urged to invest in electrochemical research and other EV-related R&D areas to keep up with the international competition.

Following this appeal, the government focused on a long-term R&D strategy instead of offering tax incentives or subsidies for consumers as France and Norway did. The latter approach would have meant encouraging consumers to buy imported EVs, as German manufacturers had no competitive electric models available when the National Initiative took off. The investment in battery R&D still seemed surprising to some experts, considering the German government and the EU had just come out of a sobering hype–disappointment cycle in hydrogen and fuel cell research.[8]

The R&D-based rearticulation of automobility constitutes the dominant narrative today. It builds on the hope that through investment in some technology the EV could be turned into a proper substitute for the conventional car. From an engineering perspective, a breakthrough such as a miracle battery or a hydrogen car represents a radical deviation from the currently dominant regime. From the motor industry's view, it would require profound changes in its production, supply chains and labour force. 'Fixing the technology' would mean 'fixing the car industry', a notion that German automakers have met with resistance. At the same time, encouraged by the state's promise to make this transition for the industry as smooth and inexpensive as possible, the National Initiative implicitly reframed the electrification of transportation as a contained, orderly transition.

The government prioritized the economic well-being of car manufacturers and its central position in any articulation of an electrified transportation future over more radical scenarios. German policymakers maintained their close relationship to the industry, while the industry, at least in theory, supported the National Initiative. The responsibility for actual progress, in terms of a large-scale diffusion of EVs, was delegated to an elaborate network of public and private R&D institutions. Since the technology was not yet ready, any serious market endeavour had to be postponed until it was. Early users validated this belief. They confirmed that consumers were, indeed, not willing to exchange their conventional car for a more expensive and somewhat limited imitation. These studies, however, took the STI of traditional automobility for granted and, thus, confirmed the inferiority of the battery car they had assumed at the outset. The imagined future in this scenario appears highly conservative. The technology fix aligns with all the key elements of the current STI of automobility. It emphasizes the modern ideals of individual self-determination and national economic progress, as well as existing infrastructures and practices such as the arrangement of private home and car ownership.

The technology fix narrative allowed the three involved groups – the state, the automotive industry and R&D institutes – to create a comfortable arrangement, which promoted the vision of electrified transportation while postponing painful changes and conflicts for the future. By reimagining only the technical components of the car as the ideal solution, all other aspects of the 'automobile nation' could be maintained and once again legitimized.

Finding a new purpose: cars as energy infrastructure

While the belief in a technological fix remains strong among experts and policymakers, the initial confidence in this type of solution has waned over time. With Germany dramatically lagging behind its goals and no technological breakthrough in sight, the media and political actors have started putting pressure on the government again. The resulting crisis has paved the way for new challengers coming from outside the transportation industry, in particular utilities, electrical grid operators and ICT companies. They promote the notion that the electrification of transportation and the transition towards renewable energy sources are two sides of the same coin. Energy systems engineers and industry experts have proposed treating the electric car as if it were part of the electrical grid infrastructure (Kempton and Tomić, 2005). Since the EV is, simply put, a large mobile battery, proponents of this approach argue that they could store surplus energy from volatile sources such as wind and solar power and feed it back into the electrical grid when needed.

Some experts argue that the integration of EVs into the 'smart grid' or 'smart home' would create synergies and open up novel economic applications (van der Kam and van Sark, 2015). The EV could be the key component needed to achieve high shares of renewable energy sources in the future energy mix (Loisel, Pasaoglu and Thiel, 2014). This technological trajectory implies a seamless interaction between the car and the electrical grid. An EV would not just charge until the battery has reached its maximum capacity; instead, the idea is that the car is actively connected to the grid as much as possible to provide technical services to either the grid operator or a local microgrid. It would be embedded within a virtual ecosystem, where it is constantly connected to a server hub, available to either the driver or another agent such as a service provider. The supporters of this scenario claim that this would raise consumer interest in EVs. If drivers made their cars available to the utilities and grid operators, they would be compensated financially, which could lower the costs of EV ownership significantly, as some economic calculations suggest (Beeton and Meyer, 2015).

In the industry and expert community, these ideas have already succeeded in revitalizing the hopes and high expectations that were disappointed in the late 2000s by low oil prices, unfavourable press and dissatisfaction with the state of battery technology. The commitment to the joint project within the industry had weakened. Investments went down as critics called the technological fix narrative into question, which had previously mitigated and unified the actor coalition.[9] While the optimism among industry experts, urban planners and policymakers slowly recovered, their attention moved further away from the market barriers to the futuristic high-tech image of mutually integrated transportation, energy and communication networks. The government reports published from 2010 to early 2014 reflect this trend (NPE, 2010b, 2011, 2012, 2014). The themes of 'energy transition' and 'information technology', which were previously marginal, now constitute an important pillar within the German electrification narrative.

I interpret the expert and policy community's search for a new purpose for the EV as an attempt to make the stagnating electrification trajectory immune to short-term economic failures. It brings new, powerful stakeholders and images into play while leaving the car industry in a pivotal position, at least for the moment. This strategy distracts from potential conflicts between, on the one hand, established manufacturers, who are used to incremental innovation and long product cycles, and, on the other hand, new players from other industries, who are interested in the potentially disruptive opening up of the locked-in system of automobility. These nascent assemblages are still very much society (and technology) 'in the making' (Callon, 1987/2012) and, thus, open to interpretation. One such understanding could be that the smart grid integration represents a technological fix somewhat similar to the miracle battery but on a larger scale. In the grander version, it is the energy system, not just private transportation, which requires a technological breakthrough to be able to become more sustainable and avoid the failures of the past.

Again, it is up to R&D consortia to elaborate the rearticulation of interconnected infrastructures. Regardless of the known issues with such systems (Mullan *et al.*, 2012), several government agencies, as well as parts of the industry, have dedicated significant efforts and funds to marrying transportation and the electrical grid. Dozens of state-supported public–private partnerships have emerged across Germany over the past couple of years, while corporate public relations departments and some individuals have boosted the idea behind these often very technical enterprises. These actors include people specializing in technology forecasting, future studies, or, in most cases, simply the promotion of early-stage products and services. Public speakers such as 'chief futurist' Lars Thomsen or Lawrence Burns, a former R&D executive at General Motors, have become vocal advocates for the radical transformation of urban landscapes. In his book *Reinventing the Automobile* (2010), Burns and other industry experts predict and encourage the radical 'transformation of the DNA' of the automobile, the electrical grid and communication networks into a 'system-of-systems' (Mitchell, Borroni-Bird and Burns, 2010, p. 7).

Even though this articulation of electrified mobility taps into the trope of 'disruptive technology' semantics championed by Silicon Valley entrepreneurs, it also resembles a much earlier mobility paradigm. In fact, the notion of private transportation and electricity as mutually integrated infrastructures is anything but new. In the late nineteenth century, densely populated cities such as New York, London and Berlin were about to be electrified. This electrification primarily concerned street lights and apartment buildings. However, many of the planning scenarios and futuristic depictions included transportation as well. At that time, the first generation of EVs had become popular among upper- and middle-class people, especially women (Mom, 2004). They viewed the 'explosion engine' as dirty, loud, inconvenient and, thus, unsuitable for an urban environment. Similar to the gasoline-fuelled myth around the 'adventure machine', which later grew into the STI of automobility, the STI of universal electrification produced its utopian images and mythologies of modernity. Berlin, for instance, sought to fashion itself into an imagined 'Elektropolis', a place where scarcity was eradicated once and for all by technological progress (Dame, 2011).

In contrast to the less ambitious hope of replacing the environmentally harmful combustion engine by fixing the seemingly deficient battery technology, the energy infrastructure narrative can be seen as a radical departure from the traditional STI of automobility. Such scenarios often place the EV explicitly within an urban environment, rather than in a rural or suburban context. Some scenarios even propose a combination of ride-sharing and, at some point, autonomous driving to maximize the efficiency and profitability of the envisioned 'system-of-systems'. Instead of aligning with the taken-for-granted image of the driver as the master of time and space, the heroic individual who expresses him or herself through the car, this rearticulation applies the logics of (electrical) grids and (communication) networks to the system of automobility. It assumes that drivers will give up their vehicles and routines to become users of mobility services offered through smart technology.

There are also much less centralized and corporate versions of this rearticulation, coming from local energy cooperatives, citizen initiatives and progressive-minded mobility experts (Canzler and Knie, 2016). They advocate independent, community-based micro smart grids, based on the same vehicle-to-grid technology. Even a single household can utilize a combination of, for example, solar panels, a stationary battery and an EV to become energy independent. Such existing products can be read as counter-narratives to the grandiose corporate smart city utopias, articulated mostly by ICT companies. Simultaneously it caters to the traditional STI of automobility, with an even stronger emphasis on bourgeois autonomy, by upholding the traditional arrangement of private home and car.

The vision of a pure patchwork of small, self-sufficient households or neighbourhoods remains marginal in the overall discourse. In most scenarios, overarching structures are meant to serve efficiency and energy security purposes. Urban planners and grid operators will most likely favour centralized, corporate-run solutions. With vehicle-to-grid assemblages being one of the more distant and elusive electrification scenarios, only time and ethnographic inquiries into the practical enactments of this rearticulation can tell how things will play out in different sociotechnical configurations.

Redefining the citizenship of mobility

The previous two rearticulations of the STI of automobility centred on technological innovation, but what if all the technology needed to electrify transportation was already available? EVs have been around for more than a century. Modern models such as GM's EV1 started out with high hopes, targeting a potential mass market. Commentators on the EV1's debacle attribute its failure not to disinterested consumers but to the catastrophic commercial strategy, which lead to the termination of existing leasing contracts and a universal recall by GM (Hård and Knie, 2001, p. 97). Reflexively engaging with past failures and shortcomings, environmentalist non-profits, citizen groups and academic experts argue that society should abandon its confidence in technological fixes and acknowledge that sociotechnical change primarily means a rearticulation of our collective desires, beliefs, and habits (Canzler and Knie, 2011). They also stress the urgency for immediate action. In their view, pollution and climate change are pressing problems that should be tackled with existing means and not postponed to the future.

Car sharing could be the first step away from the old paradigm of car ownership, at least within an urban context. In Germany, car sharing has been relatively successful in the major cities, with one million registered drivers combined across all providers (bcs, 2015). As yet, only small shares of these fleets are electric. In 2012, however, the French auto manufacturer Citroën launched the first electric-only car-sharing programme in Berlin, marketing it as 'the driving experience of the future'. Although the economic feasibility of commercially shared EVs is contested compared to conventional vehicles, the companies agree that the (perceived) limitations about the range and quick recharging do not apply in the context of car sharing. The obstacle of a high purchase price becomes irrelevant for the potential driver as well. The German motor club VCD, the eco-oriented counterpart to the General German Automobile Club, actively encourages drivers to use car sharing as often as possible, as long as the service providers are using electricity from renewable sources.

The VCD and other environmental organizations also promote mobility services beyond automobility. They campaign for the improvement of the so-called intermodality between various means of transportation. Instead of relying on cars entirely, drivers could use them as needed. Most of the time, the argument goes, cars could be left at the car park if there were other more efficient means available. Trains could allow for long-distance trips in combination with EVs for short drives to and from the station. Similarly, commuters could transfer from cars to public transport before entering the city, thus avoiding traffic jams and parking fees. The idea is not entirely new. It has been discussed by transportation experts since the 1960s, first in the context of freight transport and, increasingly, concerning public and private mobility. The vision of intermodality proposed in these scenarios often includes the EV as a substitute for the conventional car, but is not limited to automobility. Cars of any kind are more often depicted as part of the problem than part of the solution. Environmental organizations argue that within an intuitive and flexible transportation system, cars will automatically become less attractive. Within the intermodal paradigm, even the most affordable EVs with a range lower than 80 kilometres are perfectly suitable for almost every participant's mobility needs.

Unlike in the former rearticulations, there is no necessity for expensive R&D, but academic experts still play a crucial role in advancing this otherwise marginal position. Pushing the STI of automobility beyond private ownership and detaching the car from the perception of socioeconomic achievement goes against deeply rooted beliefs in Germany's modern collective identity. It also confronts the national constellation of auto manufacturers, R&D institutes and government agencies, which perceives itself to be the backbone of the German economy. As a consequence, economic and sociological studies that show the inefficiencies and failures of the system of automobility become powerful devices in public communication and political lobbying. They draw attention to little-known facts such as the rarity of long-distance trips among the vast majority of the population or the extensive periods of time cars remain idle in their car parks. Such studies have even informed the regular governmental progress reports issued by the National Platform (NPE, 2014, p. 16). Nevertheless, these figures are mostly meant to confirm the National Initiative's economic goals to be viable but obstructed by the supposedly ignorant public. The reports continuously imagine drivers who exhibit an irrational range anxiety and technological conservatism.

Although the National Initiative and its challengers both seek to convince the public of electrified transportation, they are motivated by entirely different rationales. For these mostly large German corporations represented in the National Platform, the economic stakes are high in this transition. Their rearticulations of the STI of automobility may encompass technologically radical scenarios such as turning EVs into a component of the energy infrastructure. They cannot, however, promote a vision where innovation is mostly a change of attitudes and practices. A more efficient use of cars through car sharing and the integration of public transportation into the system of automobility could mean lower sales numbers. Moreover, the STI of automobility, which helped to forge the German economy and nourished the post-war modern social imaginary, would not simply be rearticulated in terms of new technology but could fade away entirely.

In this third rearticulation, the 'citizenship of automobility' is being redefined. What the environmentalist non-profits and mobility experts are asking for is nothing less than to abolish the 'republic of drivers' and rewrite the constitution of modern mobility. The electrification of transportation is envisioned as only the beginning of a larger societal shift towards individualization, as well as sustainability, which also comprises the rapid advancement in digital communication and the so-called on demand or sharing economy. In the ideal future city, people would deliberately say 'no' to a particular type of convenience and status expression through car ownership. The advocates of this partially electrified but mostly car-free future employ similar narrative strategies, depicting cities that are highly accessible and at the same time sustainable, post-materialist and individualistic (Rammler, 2014). These themes align with an entirely different German STI, the environmentalist values and goals that date back to the early period of industrialization (Radkau, 1989). In a modern understanding, they have become institutionalized and central to public discourse after the emergence and political consolidation of the green and anti-nuclear movement in the 1970s and 1980s.

Conclusion: the dreamscapes of mobility

Throughout its history, Germany has imagined itself with and through the STI of automobility. The car industry and automobile culture paved the nation's way to modernity by capturing and placating the emerging tensions such as individualism vs. belonging, technological change vs. historical continuity, socioeconomic inclusion vs. exclusion, and rural vs. urban. In this regard, they have contributed to the constitution of what Charles Taylor calls the modern social imaginary. This interdependence has produced a robust set of practices for both engineers and drivers alike, which constitute the reality of Germany's larger economic and cultural landscape, and vice versa.

Recently and despite previous failures, the push towards an electrification of transportation has created a moment of possibility within the STI around the automobile. Although the conventional car still dominates public discourse, as well as the roads, a range of competing electrification futures has emerged within the expert and policy arena. The 2009 National Platform turned out to be mostly a paper tiger, but the issue allowed new alliances to form under the umbrella of more radical mobility futures. Each of the three presented rearticulations articulates what a desirable mobility future should look like and how and why the current regime should be changed (see Table 7.1). Heterogeneous alliances of actors co-produce and stabilize the envisioned sociotechnical configurations. As a consequence, each proposed mobility future varies in its scope, force and normative assumptions.

The rearticulations entail different notions of the involved artefacts, infrastructures and users and how they relate to each other. They align with certain aspects of the traditional image of automobility, which is in itself full of tension and ambiguity. Alternatively, they tap into STIs beyond the car, such as energy

Table 7.1 Comparing three rearticulations of the sociotechnical imaginary around the car

	'Rearticulating the old promise'	*'Finding a new purpose'*	*'Redefining the citizenship of mobility'*
Actors involved	Federal government, automotive industry, R&D stakeholders	Federal government, energy and ICT companies, futurists	Environmentalists, municipalities, mobility experts
Institutional position and power	Dominant and stable (incumbent position in the mobility field)	Emerging and unstable (challenger position in the mobility field)	Marginal; supported by a robust but lose network across society
Narrative focus	Inferiority of the EV, fixing the technology	Technological synergy and convergence	Post-materialism, behavioural change
Temporality	Breakthrough of S&T expected in near future	Long-term visions of technological utopias	Immediacy of change, urgency of action
Strategies	Continuation through partial technical innovation, postponement of radical change	Experimentation, disruption, introduction of new business models	Grass-roots advocacy, expert engagement, focus on available technology

independence, environmental protection and post-materialism. This finding applies to the attempt to redefine the 'citizenship of mobility' in particular, but also versions of the 'car as energy infrastructure' concept. Both address fundamental shifts within contemporary societies, instead of just focusing on 'fixing the technology'. The belief in some technical breakthrough remains strong. It can be seen as a hope to perpetuate the previous cultural matrix and its promises. The more radical articulations of electrified transportation challenge this belief in two contrasting ways. One situates the EV within the even greater narrative of the transition towards a sustainable society in the distant future. The other stresses the urgency of change and the fact that, to a certain degree, all the means necessary for a completely redesigned, low-emissions urban transportation network are already available.

The question is not which of these rearticulations will ultimately prevail. The past episodes in the history of German automobility – the incomplete manifestation of Reichow's 'auto-friendly city', for example – tell us that most likely the world to come will turn out to be a combination of the proposed scenarios. There is already considerable overlap and fluidity in between the narrative streams I have tried to separate neatly for the purpose of analysis. Perhaps things will play out entirely differently. Empirically probing STIs can neither point towards the most likely nor the right pathway into a (sustainable) future. Instead, this framework draws attention to the interdependence between competing technological developments, persisting path-dependencies and cross-cutting cultural underpinnings. Understanding the (re)visions of any given STI remains a goal for scholars and, hopefully, for policymakers as well. Looking at the implicit assumptions and power structures behind today's competing socio-technical configurations in the making and their blind spots from the STI perspective could help engineers, experts and policymakers to become more aware of them and open them up to debate.

Notes

1 Many thanks to Weert Canzler for his helpful comments on an earlier draft of this chapter. The chapter also benefited significantly from the remarks made by the reviewer and the editors of this volume.

2 Sociologists Hård and Knie (2000), historians Kirsch (2000), and STI scholars Fogelberg (2000); Callon (1980); and Brown (2001) have presented several case studies and analyses that looked into the reasons why electric cars have failed in such a variety of historical and national contexts. Although all studies point to similar technological and economic obstacles, no universal conclusions can be drawn for today's situation. With transportation, in particular, the opportunities to innovate seem to be closely tied to a country's or region's mobility culture and industrial politics. What has not worked in a specific time and place could very well work in a different setting – and vice versa.

3 The case study is based on continuous ethnographic fieldwork over the course of three years, including 30 interviews and informal conversations, within the German EV community. The main results for this chapter are supported by a comprehensive document analysis. My data includes public speech transcripts, observations, artefacts, government reports, commercial materials, visual advertisements and presentations at trade fairs. In particular, I looked into various next-generation application scenarios that have been proposed by companies across different sectors of industry, ranging from car manufacturers to utilities companies and IT firms. My interpretative approach is informed by a mix of sociological discourse analysis (Keller, 2012), situational analysis (Clarke, 2005) and visual sociology (Harper, 1988).

4 The most recent *Handbook of Science and Technology Studies* (2007) mentions 'cars' or the 'automobile' only a couple of times on over 1000 pages, most notably in the article on 'user–technology relationships' written by Oudshoorn and Trevor Pinch (2008).

5 My translations from the original German. The passage quoted last reads:

> Das Automobil sollte nicht nur als Symbol des Wohlstandes angesehen und bezeichnet werden. Es ist viel mehr einer der bedeutungsvollsten Aufträge an die Politik, die Voraussetzungen dafür zu schaffen, daß es für alle zum Segen des technischen Fortschritts wird und dabei seiner schönsten Aufgabe zu dienen vermag: die freundschaftlichen Beziehungen der Menschen und der Völker untereinander zu ebnen und zu vertiefen.
>
> (Sachs, 1984, p. 96)

6 BMW represents a notable exception from this inertia. The car company has seen moderate success on the market with its critically acclaimed premium 2014 models BMW i3 and i8. It also dedicated significant funds to R&D. In 2015, BMW joined forces with the market-leading electric car producer Nissan to speed up the deployment of charging infrastructure in a number of countries.

7 The rearticulations I describe stem from an inductive analysis of documents and expert interviews, which I have combined with three years of ethnographic observations in the German context. This means that there are indeed positions that are not represented here, since they have been muted or excluded from the discourse (Clarke 2005, p. 181). I decided to leave out such possible but marginal or simply unarticulated futures from

the analysis. This decision is based on the fact that sociotechnical imagination is, by definition, infinite. In order for an idea to become relevant for social practice, it must be institutionally stabilized and publicly performed, even if this consolidation is limited to a national expert community.

8 In the early 2000s, many European governments put their hopes and research funds into hydrogen research, which was, back then, believed to be close to a breakthrough (Bakker, van Lente and Meeus, 2011). Generating hydrogen from water is energy-intensive but could be done using renewable energy sources. A hydrogen-fuelled car could drive potentially carbon-free, while matching the range and refuelling speed of conventional cars. This technological fix would have combined the advantages of both the battery vehicle and the petroleum car. However, by the time the National Initiative took form, industry experts and policymakers deemed the hydrogen technology to be still too distant from a serious application and explicitly excluded it as possible scenario from the national EV road map. Recently, hydrogen has seen something of a renaissance, since it has become clear that, in spite of a rapid fall of production costs, nothing close to a miracle battery was in sight.

9 This was quite obvious when I started conducting my fieldwork in early 2012. The Forum Electromobility and the Berlin Agency for Electromobility, two state-funded governance units for the National Initiative, each hold annual conferences on the status quo of electrification. In 2012 and 2013, the atmosphere at these events can be described as pessimistic or even devastated. The discussions mainly revolved around Germany's technological lag in battery research and manufacturing, potential customers' range anxiety and the auto industry's reluctance to commit to more ambitious goals. With every subsequent year, the call for the 'miracle battery' was expressed less frequently at the conferences and workshops I participated in. Instead, application scenarios based on the idea of the EV as smart grid infrastructure gained traction in spite of their technical and economic elusiveness.

References

Anderson, B., 2006. *Imagined communities: reflections on the origin and spread of nationalism*. Rev. ed. London, New York, NY: Verso (original work published 1983).

Bakker, S. and Farla, J., 2015. Electrification of the car – will the momentum last?: Introduction to the special issue. *Environmental Innovation and Societal Transitions*, 14, pp. 1–4. doi:10.1016/j.eist.2014.07.002.

Bakker, S., van Lente, H. and Meeus, M., 2011. Arenas of expectations for hydrogen technologies. *Technological Forecasting and Social Change*, 78(1), pp. 152–162. doi:10.1016/j.techfore.2010.09.001.

Bakker, S., van Lente, H. and Meeus, M.T.H., 2012. Dominance in the prototyping phase – the case of hydrogen passenger cars. *Research Policy*, 41(5), pp. 871–883. doi:10.1016/j.respol.2012.01.007.

Ballo, I.F., 2015. Imagining energy futures: sociotechnical imaginaries of the future Smart Grid in Norway. *Energy Research & Social Science*, 9, pp. 9–20. doi:10.1016/j.erss.2015.08.015.

bcs, 2015. *Datenblatt CarSharing in Deutschland: Stand 01.01.2015*. [Online] available at: http://carsharing.de/sites/default/files/uploads/presse/pdf/datenblatt_carsharing_in_deutschland_stand_01.01.2015.pdf [accessed 12 March 2015].

Beckmann, J., 2001. Automobility: a social problem and theoretical concept. *Environment and Planning D: Society and Space*, 19(5), pp. 593–607. doi:10.1068/d222t.

Beeton, D. and Meyer, G., 2015. *Electric vehicle business models: global perspectives. Lecture notes in mobility*. Cham: Springer.

Brown, M.B., 2001. The civic shaping of technology: California's electric vehicle program. *Science, Technology & Human Values*, 26(1), pp. 56–81. doi:10.1177/016224390102600103.

Bundesregierung, 2009. *Nationaler Entwicklungsplan Elektromobilität*. Berlin. [Online] available at: www.bmbf.de/files/nationaler_entwicklungsplan_elektromobilitaet.pdf [accessed 5 August 2012].

Bundesregierung, 2011. *Regierungsprogramm Elektromobilität*. Berlin. [Online] available at: www.bmbf.de/files/programm_elektromobilitaet.pdf [accessed 5 August 2012].

Burkart, G., 1994. Individuelle Mobilität und soziale Integration. Zur Soziologie des Automobilismus. *Soziale Welt*, 45(2), pp. 216–241. doi:10.2307/40878013.

Callon, M., 1980. The state and technical innovation: a case-study of the electrical vehicle in France. *Research Policy*, 9(4), pp. 358–376.

Callon, M., 2012. Society in the making: the study of technology as a tool for sociological analysis. In: W.E. Bijker, T.P. Hughes and T. Pinch, eds. 2012. *The social construction of technological systems. New directions in the sociology and history of technology*. Cambridge, MA: MIT Press (Original work published 1987). pp. 77–97.

Canzler, W., 1999. Der anhaltende Erfolg des Automobils: zu den Modernisierungsleistungen eines außergewöhnlichen technischen Artefaktes. *Technik und Gesellschaft, Jahrbuch*, 10, pp. 19–40.

Canzler, W., 2008. The paradoxical nature of automobility. In: W. Canzler, V. Kaufmann and S. Kesselring, eds. 2008. *Transport and society. Tracing mobilities. Towards a cosmopolitan perspective*. Aldershot and Burlington, VT: Ashgate. pp. 105–118.

Canzler, W. and Knie, A., 2011. *Einfach aufladen: mit Elektromobilität in eine saubere Zukunft*. Munich: oekom.

Canzler, W. and Knie, A., 2016. Mobility in the age of digital modernity: why the private car is losing its significance, intermodal transport is winning and why digitalisation is the key. *Applied Mobilities*, 1(1), pp. 56–67.

Catenacci, M., Verdolini, E., Bosetti, V. and Fiorese, G., 2013. Going electric: expert survey on the future of battery technologies for electric vehicles. *Energy Policy*, 61, pp. 403–413. doi:10.1016/j.enpol.2013.06.078.

Clarke, A.E., 2005. *Situational analysis: grounded theory after the postmodern turn*. Thousand Oaks, CA: SAGE.

Dame, T., 2011. *Elektropolis Berlin: die Energie der Grosstadt: Bauprogramme und Aushandlungsprozesse zur öffentlichen Elektrizitätsversorgung in Berlin. Die Bauwerke und Kunstdenkmäler von Berlin*. Berlin: Gebr. Mann Verlag.

Dennis, M.A., 2015. Our monsters, ourselves: reimagining the problem of knowledge in Cold War America. In: S. Jasanoff and S.-H. Kim, eds. 2015. *Dreamscapes of modernity. Sociotechnical imaginaries and the fabrication of power*. Chicago, IL, and London: University of Chicago Press. pp. 56–78.

European Commission, 2008. *Climate change: commission welcomes final adoption of Europe's climate and energy package*. [Press release] 17 December 2008. Available at: http://europa.eu/rapid/press-release_IP-08-1998_en.htm [accessed 1 February 2014].

Featherstone, M. ed., 2004. Automobilities [Special issue]. *Theory, Culture & Society*, 21(4–5). London and Thousand Oaks, CA: SAGE.

Felt, U., 2015. Keeping technologies out: sociotechnical imaginaries and the formation of Austria's technopolitical identity. In: S. Jasanoff and S.-H. Kim, eds. 2015. *Dreamscapes of modernity. Sociotechnical imaginaries and the fabrication of power*. Chicago, IL, and London: University of Chicago Press. pp. 103–125.

Felt, U., Schumann, S. and Schwarz, C.G., 2015. (Re)assembling natures, cultures, and (nano)technologies in public engagement. *Science as Culture*, 24(4), pp. 458–483. doi: 10.1080/09505431.2015.1055720.

Fogelberg, H., 2000. *Electrifying visions: the technopolitics of electric cars in California and Sweden during the 1990s*. PhD. *STS research reports*. Gothenburg: Gothenburg University.

Franke, T. and Krems, J.F., 2013. Interacting with limited mobility resources: psychological range levels in electric vehicle use: psychology of sustainable travel behavior. *Transportation Research Part A: Policy and Practice*, 48, pp. 109–122. doi:10.1016/j.tra.2012.10.010.

Geels, F.W., 2007. Transformations of large technical systems: a multilevel analysis of the Dutch highway system (1950–2000). *Science, Technology & Human Values*, 32(2), pp. 123–149. doi:10.1177/0162243906293883.

Geels, F.W., 2012. A socio-technical analysis of low-carbon transitions: introducing the multi-level perspective into transport studies. *Journal of Transport Geography*, 24, pp. 471–482. doi:10.1016/j.jtrangeo.2012.01.021.

Hackett, E.J., Amsterdamska, O., Lynch, M. and Wajcman, J. eds., 2008. *The handbook of science and technology studies*. 3rd ed. Cambridge, MA: MIT Press; published in cooperation with the Society for the Social Studies of Science. Available at: http://eprints.lse.ac.uk/28629.

Hård, M. and Knie, A., 2000. *Getting out of the vicious traffic circle: attempts at restructuring the cultural ambience of the automobile throughout the 20th century* (No. FS II 00–103). [Online] Berlin: Wissenschaftszentrum Berlin für Sozialforschung. Available at: http://hdl.handle.net/10419/49806 [accessed 7 February 2012].

Hård, M. and Knie, A., 2001. The cultural dimension of technology management: Lessons from the history of the automobile. *Technology Analysis & Strategic Management*, 13(1), pp. 91–103. doi:10.1080/09537320120040464.

Harper, D., 1988. Visual sociology: expanding sociological vision. *The American Sociologist*, 19(1), pp. 54–70.

Hecht, G., 2009. *The radiance of France: nuclear power and national identity after World War II. inside technology*. Cambridge, MA: MIT Press.

Henderson, K., 1998. The aura of 'high tech' in a world of messy practice. *The Sociological Quarterly*, 39(4), pp. 645–672. doi:10.1111/j.1533–8525.1998.tb00522.x.

Hughes, T.P., 1983. *Networks of power: electrification in Western society, 1880–1930/ Thomas P. Hughes*. Baltimore, MD: Johns Hopkins University Press.

Hughes, T.P., 1986. The seamless web: technology, science, etcetera, etcetera. *Social Studies of Science*, 16(2), pp. 281–292. doi:10.1177/0306312786016002004.

IEA, 2009. *Transport energy and CO_2: moving toward sustainability*. Paris: OECD/IEA.

IEA, 2012. *World energy outlook 2012*. [Online] Paris: International Energy Agency. Available at: www.iea.org/publications/freepublications/publication/WEO2012_free.pdf [accessed 12 April 2013].

IEA, 2014. *Energy technology perspectives 2014: Harnessing electricity's potential/ International Energy Agency*. Paris, France: OECD/IEA.

Jasanoff, S., 2015. Future imperfect: Science, technology, and the imaginations of modernity. In: S. Jasanoff and S.-H. Kim, eds. 2015. *Dreamscapes of modernity. Sociotechnical imaginaries and the fabrication of power*. Chicago, IL, and London: University of Chicago Press. pp. 1–33.

Jasanoff, S. and Kim, S.-H., 2009. Containing the atom: sociotechnical imaginaries and nuclear power in the United States and South Korea. *Minerva*, 47(2), pp. 119–146. doi:10.1007/s11024–009–9124–4.

Jasanoff, S. and Kim, S.-H., 2013. Sociotechnical imaginaries and national energy policies. *Science as Culture*, 22(2), pp. 189–196. doi:10.1080/09505431.2013.786990.

Jasanoff, S. and Kim, S.-H. eds., 2015. *Dreamscapes of modernity: sociotechnical imaginaries and the fabrication of power*. Chicago, IL, and London: University of Chicago Press.

Keller, R., 2012. *Doing discourse research: an introduction for social scientists*. London: SAGE.

Kempton, W. and Tomić, J., 2005. Vehicle-to-grid power implementation: from stabilizing the grid to supporting large-scale renewable energy. *Journal of Power Sources*, 144(1), pp. 280–294. doi:10.1016/j.jpowsour.2004.12.022.

Kirsch, D., 2000. *The electric vehicle and the burden of history*. New Brunswick, NJ: Rutgers University Press.

Kocka, J., 1988. German history before Hitler: the debate about the German Sonderweg. *Journal of Contemporary History*, 23(1), pp. 3–16. doi:10.2307/260865.

Latour, B., 1986. Visualization and cognition: drawing things together. In: E. Long and H. Kuklick, eds. 1986. *Knowledge and society. Studies in the sociology of culture past and present*. Greenwich, CT, and London: JAI. pp. 1–40.

Loisel, R., Pasaoglu, G. and Thiel, C., 2014. Large-scale deployment of electric vehicles in Germany by 2030: an analysis of grid-to-vehicle and vehicle-to-grid concepts. *Energy Policy*, 65, pp. 432–443. doi:10.1016/j.enpol.2013.10.029.

Mayntz, R. and Hughes, T.P., 1988. *The development of large technical systems. Publications of the Max-Planck-Institut für Gesellschaftsforschung: vol. 2*. Frankfurt am Main: Campus.

McCray, P., 2012. *The visioneers: how a group of elite scientists pursued space colonies, nanotechnologies, and a limitless future*. Princeton, NJ: Princeton University Press.

Mitchell, W.J., Borroni-Bird, C. and Burns, L.D., 2010. *Reinventing the automobile: personal urban mobility for the 21st century*. Cambridge, MA: Massachusetts Institute of Technology.

Mom, G., 1999. Inventing the miracle battery: Thomas Edison and the electric vehicle. *History of Technology*, 20, pp. 17–46.

Mom, G., 2004. *The electric vehicle: technology and expectations in the automobile age*. Baltimore, MD: Johns Hopkins University Press.

Mullan, J., Harries, D., Bräunl, T. and Whitely, S., 2012. The technical, economic and commercial viability of the vehicle-to-grid concept: special section: frontiers of sustainability. *Energy Policy*, 48, pp. 394–406. doi:10.1016/j.enpol.2012.05.042.

NPE, 2010a. *Bericht der AG-2 Batterietechnologie für den Zwischenbericht der Nationalen Plattform Elektromobilität*. [Online] Berlin: NPE. Available at: www.bmvi.de/SharedDocs/DE/Anlage/VerkehrUndMobilitaet/elektromobilitaet-ag-2.pdf?__blob=publicationFile [accessed 12 May 2012].

NPE, 2010b. *Zwischenbericht der Nationalen Plattform Elektromobilität*. [Online] Berlin: NPE. Available at: www.bmub.bund.de/fileadmin/bmu-import/files/pdfs/allgemein/application/pdf/bericht_emob_bf.pdf [accessed 12 May 2012].

NPE, 2011. *Second report of the National Platform for Electromobility*. [Online] Berlin: NPE. Available at: www.bmub.bund.de/fileadmin/bmu-import/files/english/pdf/application/pdf/bericht_emob_2_en_bf.pdf [accessed 5 June 2014].

NPE, 2012. *Fortschrittsbericht der Nationalen Plattform Elektromobilität (Dritter Bericht)*. [Online] Berlin: NPE. Available at: www.bmub.bund.de/fileadmin/bmu-import/files/pdfs/allgemein/application/pdf/bericht_emob_3_bf.pdf [accessed 5 June 2014].

NPE, 2014. *Fortschrittsbericht 2014: Bilanz der Marktvorbereitung*. [Online] Berlin: NPE. Available at: www.bmwi.de/DE/Mediathek/publikationen,did=672614.html [accessed 12 August 2014].

Oudshoorn, N. and Pinch, T. eds., 2003. *Inside technology. How users matter: the co-construction of users and technologies*. Cambridge,MA: MIT Press.

Oudshoorn, N. and Pinch, T., 2007. User-technology relationships: some recent developments. In: E.J. Hackett, O. Amsterdamska, M. Lynch and J. Wajcman, eds. 2007. *The handbook of science and technology studies*. 3rd ed. Cambridge, MA: MIT Press; published in cooperation with the Society for the Social Studies of Science.

Ozaki, R., Shaw, I. and Dodgson, M., 2013. The coproduction of 'sustainability': negotiated practices and the Prius. *Science, Technology, & Human Values*, 38(4), pp. 518–541. doi:10.1177/0162243912441029.

Pinch, T.J. and Bijker, W.E., 1984. The social construction of facts and artefacts: or how the sociology of science and the sociology of technology might benefit each other. *Social Studies of Science*, 14(3), pp. 399–441. doi:10.2307/285355.

Polster, W. and Voy, K., 1991. Eigenheim und Automobil: die Zentren der Lebensweise. In: K. Voy, W. Polster and C. Thomasberger, eds. 1991. *Gesellschaftliche Transformationsprozesse und materielle Lebensweise. Beiträge zur Wirtschafts- und Gesellschaftsgeschichte der Bundesrepublik Deutschland (1949–1989)*. Marburg: Metropolis. pp. 294–356.

Radkau, J., 1989. *Technik in Deutschland: vom 18. Jahrhundert bis zur Gegenwart* (1. Aufl.). Edition Suhrkamp N.F., Bd. 536: Vol. 1536. Frankfurt am Main: Suhrkamp.

Rammert, W., 1997. New rules of sociological method: rethinking technology studies. *The British Journal of Sociology*, 48(2), pp. 171–191.

Rammert, W., 2002. The cultural shaping of technologies and the politics of technodiversity. In: K.H. Sørensen and R. Williams, eds. 2002. *Shaping technology, guiding policy. Concepts, spaces, and tools*. Cheltenham and Northampton, MA: Edward Elgar. pp. 173–196.

Rammert, W., 2012. Distributed agency and advanced technology. In: J.-H. Passoth, B.M. Peuker and M.W.J. Schillmeier, eds. 2012. *Routledge advances in sociology: vol. 58. Agency without actors? New approaches to collective action*. London and New York, NY: Routledge. pp. 98–112.

Rammler, S., 2008. The Wahlverwandtschaft of modernity and mobility. In: W. Canzler, V. Kaufmann and S. Kesselring, eds. 2008. *Transport and society. Tracing mobilities. Towards a cosmopolitan perspective*. Aldershot and Burlington, VT: Ashgate. pp. 57–76.

Rammler, S., 2014. *Schubumkehr: die Zukunft der Mobilität* (1. Aufl.). *Fischer Taschenbuch: Vol. 3079*. Frankfurt am Main: FISCHER Taschenbuch.

Reichow, H.B., 1959. *Die Autogerechte Stadt: ein Weg aus dem Verkehrs-Chaos*. Ravenaburgh: Otto Maier.

Rieger, B., 2013. *The people's car: a global history of the Volkswagen Beetle*. Cambridge, MA: Harvard University Press.

Rogers, E.M., 1983. *Diffusion of innovations*. 3rd ed. New York, NY, and London: Free Press; Collier Macmillan (Original work published 1962).

Rosner, L., 2004. *The technological fix: how people use technology to create and solve problems. Hagley perspectives on business and culture*. New York, NY: Routledge.

Sachs, W., 1984. *Die Liebe zum Automobil: Ein Rückblick in die Geschichte unserer Wünsche* (1. Aufl.). Reinbek bei Hamburg: Rowohlt.

Sachs, W., 1992. *For love of the automobile: looking back into the history of our desires*. Berkeley, CA: University of California Press.

Schwedes, O., Kettner, S. and Tiedtke, B., 2013. E-mobility in Germany: white hope for a sustainable development or fig leaf for particular interests? *Environmental Science & Policy*, 30, pp. 72–80. doi:10.1016/j.envsci.2012.10.012.

Seiler, C., 2008. *Republic of drivers: a cultural history of automobility in America*. Chicago, IL: University of Chicago Press.

Shove, E., Pantzar, M. and Watson, M., 2012. *The dynamics of social practice: everyday life and how it changes*. Los Angeles: SAGE.

Sierzchula, W., Bakker, S., Maat, K. and van Wee, B., 2014. The influence of financial incentives and other socio-economic factors on electric vehicle adoption. *Energy Policy*, 68, pp. 183–194. doi:10.1016/j.enpol.2014.01.043.

Taylor, C., 2004. *Modern social imaginaries. Public Planet Books*. Durham, NC: Duke University Press.

Urry, J., 2004. The 'system' of automobility. *Theory, Culture & Society*, 21(4–5), pp. 25–39. doi:10.1177/0263276404046059.

Urry, J., 2007. *Mobilities*. Cambridge: Polity.

van der Kam, M. and van Sark, W., 2015. Smart charging of electric vehicles with photovoltaic power and vehicle-to-grid technology in a microgrid: a case study. *Applied Energy*, 152, pp. 20–30. doi:10.1016/j.apenergy.2015.04.092.

van der Vleuten, E., 2004. Infrastructures and societal change. A view from the large technical systems field. *Technology Analysis & Strategic Management*, 16(3), pp. 395–414. doi:10.1080/0953732042000251160.

Volti, R., 2004. *Cars and culture: the life story of a technology. Greenwood technographies*. Westport, CT: Greenwood.

Wentland, A., 2016. Imagining and enacting the future of the German energy transition: electric vehicles as grid infrastructure. *Innovation: The European Journal of Social Science Research*, pp. 1–18. doi:10.1080/13511610.2016.1159946.

Part III
Shaping societies

8 Parameters of nation-ness and citizenship in Belgium (1846–1947)

Kaat Louckx

Introduction

The explosion of information and communication technologies may be considered as one of the most distinguishing features of our modern world. In the 'information society' (Castells), the creation, distribution and manipulation of statistical data make up one of the main techniques used across all sectors of society to obtain information about all kinds of societal issues, such as health care, employment, economic productivity, crime, security, natural disasters and resource management. In the 'Age of Big Data', the march of quantification sweeps gently through academia, business, industry and government. The diffusion of statistical thinking has led to a situation in which it has become nearly impossible to 'imagine' society without relying on its statistics, charts, averages and deviations.

The classification systems on which the modern data processing applications are built have a relatively long history. As Geoffrey C. Bowker and Susan Leigh Star put it in *Sorting Things Out – Classification and its Consequences*, 'It is no happenstance that a series of universal classifications emerged in the late nineteenth century' (1999, p. 132). Bowker and Starr particularly point to the need to explore the origin and the social consequences of classifications which are formalized, institutionalized and adopted in large-scale information infrastructures. As the authors point out, 'this integration began roughly in the 1850s, coming to maturity in the late nineteenth century with the flourishing of systems of standardization for international trade and epidemiology' (1999, p. 326).

While Bowker and Starr present analyses of a variety of classification systems lodged in different communities of practice, they mainly examine the 'invisibility' of *contemporary* classifications. They devote far less scholarly effort to the *longue durée history* of specific practices and methods to count and classify. In other contributions to the STS literature, too, analyses of the historical dimensions of classification systems have hitherto remained scarce (see also Müller-Wille, 2014, p. 600). In scholarly contributions to the history of science, on the other hand, not much interest has been shown in the genesis and the social consequences of classification systems that are used in our 'information society'. Historians of science have not yet paid much attention to the technologies and infrastructures within which governmental, sociopolitical practices emerge and become 'invisible' (see also Vandermoere and Vanderstraeten, 2012; Vanderstraeten and Vandermoere, 2015).

This chapter intends to contribute to the history of the forms and categories used in statistical classification systems. Building upon the STS literature, I argue that the concepts and categories, designed to standardize the basis of observation of the statisticians, are also instruments to educate the observers and the observed. They eventually end up disciplining the 'phenomena' themselves (see also Hacking, 1982; Hentsch, 2000; Jasanoff, 2004; Igo, 2007). Analysing the history of statistical constructions may thus help us to elucidate the *conditions of possibility* for the emergence of contemporary categories and classification schemes used by modern governments (Wade, 2014; Louckx, 2015). The study of the historical construction of statistical typologies and classifications may also shed light on the historical processes through which labelling options are foreclosed and the range of possibilities for new classifications are preset (Lemke, 1995).

From a historical perspective, there is a clear link between the development of statistics and the rise of nation states. As the etymology of the term indicates, statistics was originally *state-istics*, a scientific representation of the state for administrative, governmental purposes. State-istics had to cover the growing need for information in the emerging 'enlightened' political regimes in Europe. Ian Hacking has used the phrase 'avalanche of printed numbers' to characterize the rapid development of information technologies and the growing scale of statistics in the nineteenth century. For illustrative purposes, Hacking refers to the US census and its expansion. In 1790, the first census asked four questions of each household. In 1880, the tenth census posed 13,010 questions in various questionnaires addressing people, firms, farms, hospitals, churches and more. Statistical thinking has thus also been implanted in a bureaucratic machinery. Behind the rise of state-istics and the avalanche of printed numbers, as Hacking puts it, 'lay new technologies for classifying and enumerating, and new bureaucracies with the authority and continuity to deploy the technology' (1990, pp. 2–3). In this sense, we can say that population statistics is both a cause and a consequence of our predominant views of what is modern society (Holmberg, Bischof and Bauer, 2012; Louckx and Vanderstraeten, 2015).

Statistics, in other words, not only provides scientific or 'objective' representations of the state of the state. It also rationalizes and standardizes its object into administratively convenient formats. It may, moreover, be added that the state is well equipped to insist on treating individuals according to its proper classifications and designations. It is able to use and impose its own instruments in order to distinguish between individuals who belong and individuals who do not, to acquire from its population the resources it needs to survive, to exclude from among the beneficiaries of state largess those parts of the population deemed ineligible for benefits, and so on (Scott, 1998, pp. 82–83). Seen from this perspective, analysing the history of statistical constructions may help us to shed light on the 'systems of thought' underlying the categories and classification schemes used by states to enumerate, to represent, to classify and to 'take care' of their respective populations. It may help us to elucidate the ways in which the modern nation state is indeed, to adopt Benedict Anderson's (1983) felicitous phrase, an 'imagined community'. Analysing the history of the forms and categories used in state-istics may hence shed light on the politics of membership in modern nation states. In particular, it may clarify some of the 'banal' or 'invisible' ways in which states tend to take care of their populations (see also Billig, 1995, 2009; Louckx and Vanderstraeten, 2014).

This chapter presents a case study that focuses on the historical genesis of the classification systems used in a particular kind of modern technology for screening and sorting large populations, namely the modern population census. The census is expected to embrace or include the state's population. The term is commonly used 'to refer to an attempt to count all the people in a country at a given point in time' (Headrick, 2000, p. 76; see also Desrosières, 1998, 2008). In spite of this strong emphasis on social inclusion, however, census classifications also distinguish between different modes of belonging. Hereafter I will present a historical-sociological analysis of the population censuses conducted in Belgium. I believe that Belgium allows for a particularly interesting case study not only because of the specific characteristics of the Belgian nation state, which resulted in fundamental processes of state reform during the second half of the twentieth century, but also because of the fact that the Belgian *homo statisticus* Adolphe Quetelet played a key role in the development of population censuses and state-istics – both nationally and internationally. Perhaps the following case study therefore also has broader relevance. The time frame of my analysis consists of a period of about one century: from the first Belgian population census, taken in 1846, to the tenth population census, conducted in 1947. As we will see, it is in this time period that contemporary representations of the Belgian nation state and its state citizenship emerged. It is also the period within which Bowker and Starr saw the development of universal classification systems in the contemporary world.

My primary historical sources are the Belgian census reports. These reports contain a presentation of the results of the census, as well as a rather technical part that reproduces the instructions to the census-takers. I will refer to these reports by the letter B, followed by the year the census was taken. In order to contextualize these historical sources, I have also made use of reports of international statistical organizations, such as the International Statistical Congress, and of Belgian juridical sources, such as the Belgian Civil Code and the Belgian Constitution.

This chapter thus focuses on the different *parameters of nation-ness and citizenship* underlying the Belgian population censuses. As will be indicated in more detail in the following sections, three different items have been connected with the nation state as such in the Belgian censuses: language, place of birth and nationality. Hereafter I will first discuss the language question and its modifications in the course of time. While language is said to have a relatively defined territorial component in the modern era (Gellner, 1983; Rokkan, 1999), the idea of ‘one nation, one language’ was also contested in Belgium. I show how language has become the source of relatively strong social and cultural cleavages within the Belgian nation state – as exemplified by the well-known tensions between its different linguistic communities in the second half of the twentieth century. Afterwards, I analyse the ways in which data about each individual’s birthplace and nationality were collected and processed. These analyses also show how imaginaries of national identity correspond with changing state-istical interests in managing and controlling migration in a world characterized by increasing transnational mobility. In the final section, I conclude with a discussion of the historical and contemporary relevance of my analysis of parameters of nation-ness and citizenship for the project of nation-building and imagining the nation state within Belgium.

Language

In early nineteenth-century Europe, the idea had started to gain ground that language and territory were (or had to be) inextricably linked. One more or less standardized and unified language was increasingly seen as an identifying marker for a particular state; knowledge of ‘its’ language also became an essential requirement for social mobility in the state (see Burke, 2004; Edwards, 2009). Peter Burke (2004, p. 10), for example, refers to the French priest and politician Henri Grégoire, who at the time of the French Revolution started to propagate the teaching of Standard French all over France in order to ‘fondre tous les citoyens dans une masse nationale’ (melt all citizens into a national mass).

However, the nineteenth century not only saw the development of direct links between territorial and linguistic unity; in various parts of Europe, linguistic diversity also gained ground. Several ‘other’ languages used, or formerly used, within the territory of the state were able to be revived. Examples include the Frisian, Norwegian or Finnish movements. It was also in the context of this struggle between unifying forces and, to some extent, diversifying forces, that Belgium was founded in 1830.

The Southern Netherlands was united with the Northern Netherlands to form the United Kingdom of the Netherlands at the Congress of Vienna in 1814, but a range of cultural differences created obstacles for King William I’s unification policies. Religious matters (the Protestant North vs. the Catholic South) in particular were important in the conflict preceding the separation. The language issue also played a role in the conflict. William’s language policy had aimed at uniting the two regions under a common Dutch language. After the Belgian Revolution, the new constitution guaranteed freedom of language (Servais and Mechelynck, 1933, p. 5). In practice, however, French was perceived as the more prestigious language and quickly replaced Dutch in all official domains and official functions (see e.g. Dubois, 2005; Vandenbussche, 2007; Wils, 2009). French was not only the language of Enlightenment, progress and modernity; it was above all also a symbol for the national struggle for independence (Vandenbussche, 2007).

Until 1930, the census reports were drafted solely in French. With the exception of the 1856 census, all censuses taken between 1846 and 1947 included language items. In the first Belgian census, the statisticians defined language as one of the most enduring distinguishing features of people or populations.[1] The census included, more particularly, a question on the *habitually* spoken language (*la langue parlée habituellement*) (B 1846, LV). People were asked to declare the language they *most frequently* used.[2] No options were specified; the respondents could fill in any language. In the summary tables with the results, the statisticians grouped the data in the following categories: French or Walloon (which they considered to be a variety of Standard French), Flemish or Hollandish (which was considered to be a variety of Flemish), German, English, and other languages (B 1846, XXXVII). The figures presented made clear that the majority of the Belgian population spoke either French or Flemish, while German was mainly spoken in parts of the Belgian province of Luxembourg. The statisticians added that all other languages were only used by a small group of migrant workers and merchants, and by foreigners who were only temporarily residing in Belgium (B 1846, XXXVII).

In the 1846 census report, the Belgian statisticians also invoked the authority of 'philologists' to refer to the unifying force of a common, standardized language. In their view, the line of demarcation between the French- and Flemish-speaking populations was a symptom of more fundamental social divisions within the Belgian state (*le peuple belge se trouve divisé sous le rapport du langage parlé*) (B 1846, XXXVI).

Without any explanation, the 1856 census questionnaire included no item about language. But, as of 1866, new attempts were made to inquire into the linguistic identity of the population of the Belgian nation state. The statisticians also rephrased the language item. In 1866, the aim was no longer to collect data about the *variety* of languages spoken within the country. Instead the statisticians started to collect information about the *ability* of citizens to speak French, Flemish or German. The statisticians seemed to distance themselves from the idea of 'one state, one nation, one language'. They now depicted the most commonly spoken languages in Belgium – French or Walloon, Flemish or Hollandish, (Lower) German or Luxembourgish – as the state's *national languages* (see, e.g. B 1866, XII; B 1880, XLIX; B 1890, CL; B 1900, CLXX).

From 1866 onwards, individuals were more particularly asked to declare *which* and *how many* of the Belgian national languages they habitually used. Hence, the language data could be used to determine not only the strengths of the different monolingual blocs within the state; the statisticians were clearly also starting to show interest in the incidence of bi- and trilingualism among the Belgian population (see B 1866, XXI–XXII; B 1880, XXV; B 1890, XXXV–XXXVIII; B 1900, XXXIX–XLV; B 1910, p. 105, p. 203; B 1920, pp. 16–17). It might be assumed that this interest was politically and state-istically motivated. The emphasis on bi- and trilingualism could serve to distinguish Belgium from France, with its assimilatory French-only policy, which ruled out all regional languages. This official 'presentation of self' could strengthen the Belgian identity and protect the nation state against France's expansion plans (see Vogl and Hüning, 2010; Hüning, 2013).

At the moment that the statisticians started focusing on the ability to speak one of the Belgian national languages, they also started to refrain from processing data on the use or knowledge of 'foreign' languages, such as English, Italian or Latin. People unable to speak any of the national languages also did not have to specify the languages they were able to speak; they were simply classified in the residual category of 'inhabitants who don't speak one of the national languages' (see, e.g. B 1866, XLII; B 1880, LXXXV; B 1890, CXXII; B 1900, CLVI; B 1910, p. 105).

As of 1910, while continuing to ask about the national languages spoken, the Belgian census-takers also asked those who named more than one language a second question, namely, which one they used most frequently (see, e.g. B 1910, p. 4; B 1920, p. 26; B 1930, T1, p. 41; B 1947, p. 70). In summary tables about the linguistic state of the nation, bi- or trilingual inhabitants were counted two or three times. For example, Flemish-speaking inhabitants who also had knowledge of French were added to both the Flemish- and the French-speaking populations (e.g. B 1910, p. 203; B 1920, p. 63; B 1930, T2, p. 14). But much opposition crystallized around this strategy, especially in the Flemish-speaking part of the country. As Flemings were more likely to be bilingual than those who had French as their mother tongue (among other reasons because secondary education took place solely in French on the entire Belgian territory until the late nineteenth century), the census data were believed to overestimate the number of French-speakers (see Levy, 1960; Verdoodt, 1983). Although by the end of the nineteenth century new legislation had started to grant more linguistic rights to the Flemish-speaking population, the so-called Flemish movement increasingly pursued the idea of a linguistically homogeneous Flanders. In this context, bilingualism was viewed as a transitional stage from Flemish- to French-speaking (see also O'Neill, 2000; Haarman, 2012).

By the 1880 census, the statisticians had already started to express their doubts about the completeness and correctness of the language count (B 1880, LXXXIII). The census-takers were instructed to double-check the responses to the language item (B 1890, CXXVIII–CXXX). Fines were imposed to penalize individuals who attempted to make false declarations (B 1890, CXI, CXXIX; B 1900, CXXXVII; B 1910, p. 88). These measures were taken to avoid the language questionnaire becoming used by Flemish-speaking population members to create their own (imagined) linguistic community. However, the statisticians did not believe that these measures had the hoped-for results. Flemish-speaking population members kept on refusing to declare being bilingual in the knowledge that they would be added to *both* the Flemish- and French-speaking population. In the early twentieth century, the statisticians explicitly invoked the increase of false language declarations to account for the declining number of bilingual Flemings (e.g. B 1910, pp. 203–204; B 1920, p. 64). These problems encountered with the collection of linguistic census data show how the history of statistics is also a history of contrasting and conflicting imaginaries.

Some more systematic observations may be added. Despite the focus on the *langues parlées* (the spoken languages), the census reports show many traces of the impact of standard written languages on national identity constructions in Belgium. The 'nationalization' of the languages commonly used in Belgium depended on processes of *standardization* and *homogenization*. In 1880, the census-takers were explicitly directed to consider Walloon as a variety of French, Dutch or Hollandic as varieties of Flemish, and Lower German or Luxembourgish as varieties of German (see, e.g. B 1880, LXXXV; B 1890, XXXV; B 1910, p. 44, p. 105).[3] From the 1890 census onwards, only French, Flemish and German appeared as headings in the census reports (see, e.g. B 1890, LXXXIII, LXXXV; B 1900, CV; B 1910, p. 45).

By the early twentieth century, the Flemish movement had become more oriented towards the Netherlands. The identification of the Flemish varieties in Belgium with the Dutch spoken in the Netherlands was a way to enhance the prestige and thus strengthen the position of Dutch (Flemish) against French in Belgium (Vogl and Hüning, 2010, p. 238). In 1947, the label 'Flemish' was replaced by that of Dutch in the Belgian population census (e.g. B 1947, pp. 106–112). In the eyes of the state and its state-istics, Flemish was now a variety of Dutch (while the Romance varieties spoken in Wallonia had always been viewed as varieties of Standard French). In 1932, new language legislation specified that the results of the language count had to be used to determine the language regime of the local municipalities (see B 1947, p. 52). In Flanders, however, the results of the language count (which pointed to the 'Frenchification' of Brussels and a number of other municipalities) were again highly contested. As result of this conflict, the language results of the 1947 census were only partially published. While the objectivity of the language census was called into question, pressure was also built for the establishment of a constitutional language border between the Dutch- and French-speaking regions within Belgium. Many local authorities in Flanders opposed any new language census. Owing to this political conflict, the government postponed the 1957 census (see e.g. Levy, 1960; Louckx, 1982; Verdoodt, 1983). The next census was conducted in 1961, but language questions were now dropped from the census. Shortly afterwards (1962–1963), the territorial language border was established.

In the course of the nineteenth and twentieth century, different imaginaries of the nation state have developed within Belgium. As in most other nation states, in the Belgian independent state a common and uniform language was regarded as an identifying marker that could hold all the inhabitants of the state together (see e.g. Arel, 2002; Edwards, 2009; Malesevic, 2013). Over time, however, the Belgian nation state has become subdivided into separate language territories. The official demarcation of these language territories was not simply a solution to a language conflict; it also followed from and reinforced the nineteenth-century idea of 'one state, one nation, one language'. The history of the Belgian language census and its representations of the nation state sheds light on the identity problems with which the Belgian state was faced. As we have seen, these state-istics did not simply provide information about the state of the nation state; they also created a dynamic that came to threaten the very existence of the Belgian nation state.

Place of birth and nationality

The Belgian nation state originated at a time when states were ideally linked to one language and when languages had – at least in theory – to be standardized and uniform and to serve as identifying markers holding all the members of the nation state together. But it was not only linguistic unity that was at that time linked with national unity. The concept of citizenship also arose together with the concept of the nation state. In the course of the nineteenth century, the rights conferred upon citizens started to grow in number and substance. But these rights were also increasingly confined to nationals. An increasing delimitation of national citizenries took place. Citizenship came to mean membership of the nation state (see Hobsbawn, 1990; Brubaker, 1992; Torpey, 2000).

After the Belgian Revolution of 1830, the Belgian Civil Code and the Belgian Constitution established Belgian citizenship.[4] On the basis of these laws, citizenship could be automatically attributed (by the state to the individual), acquired (by the individual through procedures prescribed by the state) and withdrawn (from the individual by the state authorities). Notwithstanding several adaptations, the *core* principles of this legislation remained the same until the early twentieth century. After the First World War, protectionism prevailed and greater restrictions in granting Belgian citizenship to foreigners were imposed (Foblets, Yanasmayan and Wautelet, 2013).

Belgian citizenship was in principle granted at birth. The dominant Belgian practice built upon the principle of *jus sanguinis paterni* (right of blood in the paternal line): a child whose father had Belgian citizenship automatically acquired Belgian citizenship, irrespective of the place of birth.[5] This mode of attribution was legitimized by the conviction that citizenship could not simply follow from 'accidentally' having been born in Belgium but had to be seen as the consequence of family or blood ties, as the heritage of a community, which is made up of people who together form a sovereign nation. But the principle of *jus soli* (right of soil) was also introduced: a child born in Belgium to non-Belgian parents, for example, could acquire Belgian citizenship if certain requirements (mainly pertaining to age and residence) were fulfilled.[6] Belgian citizenship could moreover be lost by individuals who voluntary acquired the nationality of a foreign state, voluntarily served in a foreign army, or settled abroad and did not show any evidence of an intention (*animus*) to return to Belgium.[7]

However, despite the elaboration of the politico-legal institution of Belgian citizenship, and despite its symbolic importance for the new state, the Belgian census did not include any question about legal nationality until the end of the nineteenth century (1890). In the first censuses, the census-takers were even explicitly directed *not* to take legal nationality into account. The census-takers were instructed to gather data about all those present in the household – *quelle que soit leur nationalité* (B 1866, XXXI). The Belgian statisticians used another variable to provide a more detailed overview of the modes of membership in the nation state. Instead of legal nationality, they focused on people's *place of origin or birth* (e.g. B 1846, XXXVI–XXXVII; B 1856, XIX, XLVIII, XLIX; B 1866, p. 19; B 1880, XXI; B 1890, XXIII–XXIX).

In the view of the Belgian statisticians, the question about the place of origin or birth enabled them to monitor people's 'movements' on the territory of the Belgian state (B 1846, XXXV; see also B 1856, XVIII; B 1866, XX). Initially, the statisticians showed a strong interest in both intra- and international movements (B 1846, XXV; B 1856, XIX; B 1866, p. 18; B 1880, XXII). But, in the last decades of the nineteenth century, their state-istical interest shifted almost exclusively towards *international migration*. At that time, they also started to speculate on the motives that the growing number of migrants might have. In their opinion, economic motives dominated. The statisticians argued that most *étrangers* living on Belgian territory had left their country of origin to find fortune in Belgium. They added that it was not likely that these *étrangers* would return to their country of origin or birth (e.g. B 1890, XXXIII; B 1900, XXXVII).[8]

As of 1880, the census reports started to include longitudinal overviews of population movements to and from Belgium. The table drafted by the statisticians divided the growing number of foreigners residing on Belgian territory on the basis of their place of birth, but the table's title promised an overview *par nationalité* (B 1880, XXIII). Based on census data from a number of other countries, such as France, Germany, the United States and Great Britain, the 1880 census report also tried to provide an overview of the number of 'Belgians' living abroad. But in an explanatory note, the statisticians clarified that the term 'Belgians' referred solely to people born in Belgium and *not* to people with the Belgian nationality in the legal sense of the word (B 1880, XXIV).[9] Although the law primarily distinguished between citizens and non-citizens on the basis of blood ties, the official state-istics show how *territorial* criteria predominated over legal criteria in distinguishing between nationals and foreigners. The Belgian nation state was imagined as a community of birth, of native-born citizens.

Six decades after Belgium's independence, the 1890 census was the first to include an item on the *legal* nationality status of the inhabitants. According to the statisticians, the inclusion of this item constituted *un intérêt national et international*; they also mentioned that they followed the example of several other states (B 1890, IV–V).[10] In the census report, the statisticians now made clear that only individuals who possessed Belgian nationality status could be identified as Belgians or nationals. Inversely, foreigners were defined as individuals who did not possess Belgian nationality status (B 1890, LXXXIII; B 1910, p. 49). The item on the nationality status did not replace but supplemented that on people's place of birth or origin. Different imaginaries of belonging were thus introduced. The difference was also stressed on many occasions; in the instructions to the census-takers, the statisticians repeatedly insisted that nationality declarations needed to be made in conformance with the *legal* definition of nationality (e.g. B 1900, LXXXIV; B 1910, p. 29).

The presentation of the data on legal nationality and place of birth in the census reports shows the lasting concern of the statisticians with the increase of the number of migrants. In fact, the Belgian statisticians often presented tables in which the data on nationality and place of birth were compared or crossed. In 1890, for example, the statisticians started to focus on the countries of birth and of nationality of the *largest* number of foreigners on Belgian territory (e.g. B 1890, XXXV; see also B 1900, XXV; B 1910, p. 192; B 1920, p. 59). As of 1900, they also focused on the *balance* between migration to and from Belgium (e.g. B 1900, XV; B 1910, p. 183; B 1930, T3, p. 53). It was for example noted and applauded in the 1900 census report that the emigration was higher than the immigration. While the growth of the population living on Belgian territory was mainly 'homemade', Belgium was also able to contribute considerably to the so-called 'international circulation of humanity' (B 1900, XV).[11]

As of 1890, the statisticians also devoted substantial parts of the census reports to the presentation and interpretation of cross-tabulations in which *different citizenship categories* appeared. They distinguished, for example, between Belgians born in Belgium, Belgians born abroad, individuals of foreign nationality residing in Belgium and born in Belgium, and foreigners residing in Belgium but born outside of Belgium. They not only compared the size of the different citizenship categories but also discussed changes over time (or the lack thereof) and speculated about push and pull factors, such as economic opportunities and urbanization (e.g. B 1890, XXXIV; B 1900, XXXVI–XXXVIII; B 1910, pp. 195–198; B 1920, p. 59; B 1930, T3, p. 48, pp. 54–55, pp. 58–59, p. 72). Indirectly, these tables allowed for the identification of individuals who might qualify for naturalization procedures.[12]

In the same context, the statisticians also started to pay attention to gender differences (e.g. B 1900, XXXV; B 1910, p. 195; B 1920, p. 59). For example, they linked the finding that most immigrants were men with economic factors: male immigrants were depicted as economic players, who migrated to Belgium to find jobs (B 1890, XXXIII; B 1900, XXXVII; B 1910, p. 197). The high number of women from neighbouring countries living in Belgium was attributed to the demand for household workers in Belgium (B 1890, XXXIII, XXXV; B 1910, p. 195; B 1920, p. 59; B 1930, T3, p. 77). The high proportion of women born in Belgium to foreign parents was said to be the counterpart of men's higher mobility. Men migrated more easily not only for economic reasons; they often also had to return to their country of legal nationality to fulfil their military service (B 1900, XXXVII; B 1910, p. 197). The higher proportion of women among the Belgians born abroad was linked with the nationality legislation: women born to foreign parents could become Belgian by marrying a Belgian, while men could less easily lose their foreign nationality (B 1900, XXXVII; B 1910, p. 197).

In the interwar period, the Belgian statisticians continued to publish analyses of citizenship categories in relation to migratory movements.[13] Foreigners now also needed to inform about the duration of their stay in Belgium; they also had to specify whether or not they had applied for a residence permit – thus showing their intention (or lack thereof) to stay in Belgium (see, e.g. B 1930, T1, p. 41, p. 47, p. 49; B 1947, pp. 105–106, pp. 111–112). Interestingly, the statisticians also (again) developed an interest in intra-national migration movements. In the

1930 census, more particularly, the statisticians started to calculate emigration and immigration flows between Flanders and Wallonia (B 1930, T3, p. 51). They took this internal territorial division as the point of departure of their analyses at the moment that the tensions between Flemings and Walloons were running high but well before the establishment of the language border in Belgium and the regionalization of the nation state (B 1930, T3, pp. 50–52).

After the Second World War, the bureaucratic regulations about citizenship increased once more (see also Brubaker, 1992; Simon, 1999). In the 1947 census, the statisticians asked naturalized Belgians to specify how they had acquired Belgian citizenship (B 1947, p. 51, p. 99, p. 106). The foreigners residing on Belgian territory on census day had to provide detailed information about their residence rights (B 1947, p. 67). The Second World War and the collaboration with Nazi Germany also left its traces. Individuals from whom Belgian citizenship status had been withdrawn (and who had hence become foreigners) had to be identified separately (see B 1947, p. 69). Altogether, the statisticians now distinguished between no fewer than 13 categories of Belgian citizens, seven categories of foreigners and four categories of stateless individuals (B 1947, p. 93).

This evolution reflects the rise of what Max Weber has called the 'rational-legal authority' within the modern state. With Michel Foucault, we might also speak of the *étatisation* and nationalization of membership (Foucault, 1984, pp. 302–303; see also Brubaker, 2010). In the nineteenth and twentieth centuries, citizenship became increasingly articulated as membership of nation states – in Belgium as well as elsewhere. But the history of the statistical articulation of citizenship also shows the growing grasp of the state bureaucracy. It shows how classifications have become integrated into large-scale bureaucratic infrastructures (see also Bowker and Starr, 1999). In twentieth-century Belgium, the political community became divided into ever more categories. For the modern nation state, such distinctions between citizens and citizenship categories might fulfil another set of instrumental purposes: Although citizenship is still an important factor in defining national identity, it has increasingly become a key instrument for controlling access to particular social rights, for disciplining the national citizens and for monitoring migration in a world characterized by increasing transnational mobility.

Conclusion

Until now, STS studies have mainly focused on contemporary classification systems and methods. Not much attention has hitherto been paid to the genesis of these classifications. As Bowker and Starr (1999, p. 126) put it: 'modern information technology … bears traces of its past, inheriting the inertia of the installed base upon which it was built'. In this chapter, I have tried to contribute to the largely neglected aspect of the 'history of the present' (Foucault) by focusing on the conditions of possibility for the emergence of contemporary practices and methods of counting and classifying. In particular, my historical and epistemological analyses have focused on the genesis of parameters of nation-ness and belonging underlying the Belgian population censuses. In various ways, my analyses have shed light on the creation of 'imagined communities' in the Belgian nation state. They not only displayed the various ways in which the Belgian nation state has tried to 'shape' national unity. They have also shown how the census technology not only reinforced the creation of particular imaginaries, but also actively resisted other imaginaries. In many ways, the history of the *parameters of nation-ness and citizenship* used in the Belgian censuses is at the same time a history of the struggles over belonging in and to the nation state. It is at the same time a history of conflicting and contrasting imaginaries.

Belgium has never been a state based on the principle of 'one state, one nation, one language' – although it was founded at a time when one state was ideally linked to one nation and one language. Its constitution guaranteed a number of freedoms, including the freedom of language choice, but French remained the socially and culturally dominant language throughout the nineteenth and early twentieth century. In the nineteenth and twentieth centuries, however, the Flemish movement was also able to create its 'imagined community'. It built on the same nineteenth-century principle of 'one state, one nation, one language'. In the second half of the last century, the pursuit of a linguistically homogeneous Flemish state has become part of Flemish mainstream politics. As I have pointed out, conflicts over the state-istical representations of language played an important role in the course of the historical process. Analyses of the various ways in which linguistic unity and national identity were counted and imagined also shed light on the *premises* and *principles* underlying the Belgian nation state. In this setting, what was not counted was as important as what was counted.

My analyses have shown how in the course of the nineteenth and twentieth centuries different imaginaries of nation-ness and belonging developed within Belgium. They also reveal shifting expectations regarding membership of the Belgian nation state. The many changes in the statistical methodologies and nomenclatures were not only the result of technical or 'material' improvements in the state bureaucracies. The changes in the items included in the census questionnaires also reflected changes in the kind of information that was of interest to the state and changes in the way in which the state defined its own function or position vis-à-vis its population. The historical variability in statistical nomenclatures and methodologies thus refers not merely to changes in statistical tools and techniques but also to statistical systems of thought and governmental practices.

As my analyses have shown, the statisticians primarily distinguished between nationals and foreigners on the basis of territorial criteria in the first Belgian censuses. They initially imagined the nation as a community of native-born citizens. At the end of the nineteenth century, different imaginaries of belonging were introduced. In the 1890 census, they also started to record people's legal nationality statuses. Over time, more and more elaborate and complex classifications of people's citizenship status were put to use.

The census reports show many traces of the impact of processes of *standardization* and *homogenization* on national identity constructions in Belgium. My analyses have also shown how imaginaries of national identity correspond with changing state-istical interests in managing and controlling migration in a world characterized by increasing transnational mobility. As we have seen, this historical shift evolved with and reflected the development of a new politics of belonging, which is characterized by the *étatisation* and nationalization of membership or citizenship. The changes in the ways in which statisticians distinguished between citizens and non-citizens were obviously related with the development of regulatory, bureaucratic systems that control the attribution of membership rights to individuals. The quest for governmental control transformed the Belgian nation state and its state-istics by placing a premium on the articulation of the relation between citizen, territory and the nation state. I hope this case study demonstrates the usefulness of examining the historical genesis of 'banal' distinctions in the contemporary world. I hope it illustrates how current imaginaries of the future do draw upon past imaginaries of the future.

Notes

1 'Parmi les caractères qui distinguent les populations entre elles, l'un des plus inaltérables est sans contredit le langage' (B 1846, XXXVI). In the second half of the nineteenth century, similar ideas prevailed at the level of the International Statistical Congress. Language was referred to as the *core marker* of nationality; concomitantly, language items had to be used to obtain information about the different nationalities within a state (e.g. Commission Permanente du Congrès International de Statistique, 1874, p. 11, pp. 37–38, p. 43).

2 The census-takers were required to be 'intelligent men with perfect knowledge of the area where they were active' (B 1866, VIII). As such, they were often recruited among 'trustworthy' professionals, such as schoolteachers, policemen and municipal servants. A couple of weeks after the distribution of the enumeration forms, the census-takers travelled to designated households to collect the census forms. They also assisted residents by reading forms and answering questions about the census. This powerful position of 'translator' made them important persons of trust, certainly for the numerous illiterate (see also Bracke, 2008, p. 313).

3 Until the census of 1930, the census-takers were solely required to have knowledge of the official (namely, most commonly spoken) language of their locality (B 1846, L; B 1856, XXXVIII). Until that time, the census questionnaires were also distributed in the language most commonly spoken in the municipality. Because of the minority position of Flemish, Flemish-speaking inhabitants not only often received a French questionnaire, but also often didn't speak the same language as their census-taker. As most census-takers were French-speaking, some Flemish activists feared that they would register Flemish-speaking population members as French-speaking, making the country 'more French' in statistics than it was in reality (Hannes and Schmook, 1981).

4 In the first following paragraphs, the term 'citizenship' is used as synonymous with 'legal nationality'. The term refers here specifically to the official right to belong to a particular country.

5 Nowadays, the *jus sanguinis* principle is still a key element in Belgian nationality law.

6 The principle of 'one family, one citizenship' was also stressed. For instance, a foreign woman who married a Belgian man took the citizenship of her husband and so did the children born in wedlock.

7 By 1914 voluntarily serving in a foreign army was no longer a reason to revoke Belgian nationality.

8 In the last decades of the nineteenth century, the census-takers also started to record more detailed information about the place of origin of foreigners. Foreigners no longer only had to inform about the country in which they were born, but also had to provide more specific information, such as municipality and province of birth (see B 1890, IV-V, LXXXIII; B 1900, CLVII; B 1910, p. 43, p. 107; B 1920, p. 21; B 1930, T1, p. 40; B 1947, p. 110).

9 The footnote to the table reads as follows: 'Le mot Belge doit être ici entendu seulement en ce sens qu'il s'agit de personnes nées en Belgique, abstraction faite de la question de nationalité légale' (B 1880, XXIV).

10 As of 1880, the Belgian census reports show evidence of growing international collaboration among statisticians and other state servants (see B 1880, VII, VIII; B 1890, V). The efforts made by nation states to process and share legal nationality data may be regarded as indicative of the growing importance of legal nationality as a marker of belonging to the nation state.

11

> Il est très intéressant de noter que la Belgique ne tire que de ses propres forces l'augmentation remarquable de sa population et que … elle fournit encore un contingent d'environ 3 par mille de ses habitants à la circulation internationale de l'humanité.
>
> (B 1900, XV)

12 The Belgian Constitution distinguished between two types of naturalization, namely the ordinary naturalization and the full or state naturalization. Only individuals who gained state naturalization were allowed full political rights. Throughout the nineteenth century, both types of naturalization were rarely granted. Both procedures were also costly and complicated. To be eligible for ordinary naturalization, an individual needed to be at least 21 years of age and to have resided in Belgium for at least five years. State naturalization was only granted to those who had rendered 'special services to the country'. The law of 6 August 1881 made it easier for individuals who were born abroad to gain state naturalization.

13 It should, however, be added that, in 1920, thus shortly after the First World War, it was only possible to organize a limited census. In 1930, the statisticians also indicated that it became difficult to establish and interpret the longitudinal data series, because the territorial borders of Belgium were redrawn after the First World War (B 1930, T3, p. 74). Moreover, the German-speaking 'Eastern cantons' came to Belgium as a consequence of the 1919 Versailles Treaty.

References

Belgian census reports

B 1846 *Population. Recensement Général (15 Octobre 1846)*. Brussels: Le Ministre de l'Intérieur, 1849.

B 1856 *Population. Recensement Général (31 Décembre 1856)*. Brussels: Le Ministre de l'Intérieur, 1861.

B 1866 *Population. Recensement Général (31 Décembre 1866)*. Brussels: Le Ministre de l'Intérieur, 1870.

B 1880 *Population. Recensement Général (31 Décembre 1880)*. Brussels: Le Ministre de l'Intérieur, 1884.

B 1890 *Population. Recensement Général de 1890. Tome I*. Brussels: Le Ministre de l'Intérieur et de l'Instruction Publique, 1893.

B 1900 *Population. Recensement Général du 31 Décembre 1900. Tome I.* Brussels: Le Ministre de l'Intérieur et de l'Instruction Publique, 1903.

B 1910 *Recensement Général de la Population du 31 décembre 1910. Tome I.* Brussels: Le Ministre de l'Intérieur, 1916.

B 1920 *Population. Recensement Général du 31 Décembre 1920. Tome I.* Brussels: Le Ministre de l'Intérieur et de l'Hygiène, 1926.

B 1930 *Population. Recensement Général au 31 Décembre 1930. Tome I.* Brussels: Ministère de l'Intérieur, 1934.

B 1930 *Population. Recensement Général au 31 Décembre 1930. Tome II.* Brussels: Ministère de l'Intérieur, 1936.

B 1930 *Population. Recensement Général au 31 Décembre 1930. Tome III.* Brussels: Ministère de l'Intérieur, 1937.

B 1947 *Algemene Volks-, Nijverheids- en Handelstelling op 31 December 1947. Deel 1.* Brussels: Ministerie van Economische Zaken, 1949.

Other references

Anderson, B., 1983. *Imagined communities: reflections on the origin and spread of nationalism.* London: Verso.

Arel, D., 2002. Language categories in censuses: backward- or forward-looking. In: D.I. Kertzer and D. Arel, eds. 2002. *Census and identity: the politics of race, ethnicity, and language in national censuses*. Cambridge: Cambridge University Press. pp. 92–120.

Billig, M., 1995. *Banal nationalism*. London: SAGE.

Billig, M., 2009. Reflecting on a critical engagement with banal nationalism – reply to Skey. *Sociological Review*, 57(2), pp. 347–352.

Bowker, G.C. and Star, S.L., 1999. *Sorting things out: classification and its consequences*. Cambridge, MA: MIT Press.

Bracke, N., 2008. *Een monument voor het land. Overheidsstatistiek in België, 1795–1870.* Ghent: Academia.

Brubaker, R., 1992. *Citizenship and nationhood in France and Germany*. Cambridge, MA: Harvard University Press.

Brubaker, R., 2010. Migration and membership. *Journal of Interdisciplinary History*, 41(1), pp. 61–78.

Burke, P., 2004. *Languages and communities in early modern Europe*. Cambridge: Cambridge University Press.

Commission Permanente du Congrès International de Statistique, 1874. *Compte-rendu des conférences de Stockholm en 1874*. Saint Petersburg: Trenké & Fusnot.

Desrosières, A., 1998. *The politics of large numbers*. Cambridge, MA: Harvard University Press.

Desrosières, A., 2008. *Gouverner par les nombres: L'argument statistique II.* Paris: Presses de l'École des Mines.

Dubois, S., 2005. *L'invention de la Belgique: génèse d'un état-nation (1648–1830)*. Brussels: Racine.

Edwards, J., 2009. *Language and identity*. Cambridge: Cambridge University Press.

Foblets, M.-C, Yanasmayan, Z. and Wautelet, P., 2013. *Country report: Belgium*. Florence: EUDO Citizenship Observatory.

Foucault, M., 1984. Le pouvoir, comment s'exerce-t-il? In: H. Dreyfus and P. Rabinow, eds. 1984. *Michel Foucault: un parcours philosophique*. Paris: Gallimard. pp. 297–308.

Gellner, E., 1983. *Nations and nationalism*. Oxford: Oxford University Press.

Haarman, H., 2012. Language and ethnicity in a European context. In: M. Hüning, U. Vogl and O. Moliner, eds. 2012. *Standard languages and multilingualism in European history, vol. 1*. Amsterdam: John Benjamins. pp. 97–126.
Hacking, I., 1982. Biopower and the avalanche of printed numbers. *Humanities in Society*, 5, pp. 279–295.
Hacking, I., 1990. *The taming of chance*. Cambridge: Cambridge University Press.
Hannes, J. and Schmook, G., 1981. De volkstelling van 1846 en een protestbrief van H. Conscience, *Tijdschrift van de VUB*, XII, pp. 41–61.
Headrick, D.R., 2000. *When information came of age: technologies of knowledge in the age of reason and revolution, 1700–1850*. Oxford: Oxford University Press.
Hentsch, T., 2000. Compter et conter: Le dire de la statistique. In: J.-P. Beaud and J.-G. Prévost, eds. 2000. *L'ère du chiffre. Systèmes statistiques et traditions nationales*. Québec: Presses de l'Université du Québec. pp. 483–486.
Hobsbawn, E.J., 1990. *Nations and nationalism since 1780: programme, myth, reality*. Cambridge: Cambridge University Press.
Holmberg, C., Bischof, C. and Bauer, S., 2012. Making predictions: computing populations. *Science, Technology, & Human Values*, 38(3), pp. 398–420.
Hüning, M., 2013. Standardsprachenideologie. Über Sprache als Mittel zur Ab- und Ausgrenzung. In: E. Besamusca, C. Hermann and U. Vogl, eds. 2013. *Out of the box: Über den Wert des Grenzwertigen*. Vienna: Praesens. pp. 105–122.
Igo, S.E., 2007. *The averaged American. Surveys, citizens, and the making of a mass public*. Cambridge: Harvard University Press.
Jasanoff, S., 2004. *States of knowledge: the co-production of science and the social order*. New York, NY: Routledge.
Lemke, J., 1995. *Textual politics: discourse and social dynamics*. London and Bristol, PA: Taylor & Francis.
Levy, P.M., 1960. *La querelle du recensement*. Brussels: Institut Belge de Science Politique.
Louckx, F., 1982. *Vlamingen tussen Vlaanderen en Wallonië: taalaanvaardings- en taalontwijkingsprocessen in een meertalige situatie, bekeken vanuit de sociologische literatuur over etnische en raciale verhoudingen (Reeks Taal en Sociale Integratie, Volume V)*. Brussels: Vrije Universiteit Brussel.
Louckx, K., 2015. *Statistics or state-istics? An anatomy of the corps social presented in the Belgian population censuses (1846–1947)*. PhD. Ghent University and the University of Antwerp.
Louckx, K. and Vanderstraeten, R., 2014. State-istics and statistics: exclusion categories in the population census (Belgium, 1846–1930). *The Sociological Review*, 62(3), pp. 530–546.
Louckx, K. and Vanderstraeten, R., 2015. Household and state-istics: cornerstones of society in population censuses (Belgium, 1846–1947). *Social Science History*, 39(2), pp. 201–215.
Malesevic, S., 2013. *Nation-states and nationalisms: organization, ideology and solidarity*. Cambridge: Polity.
Müller-Wille, S., 2014. Race and history: comments from an epistemological point of view. *Science, Technology, & Human Values*, 39(4), pp. 597–606.
O'Neill, M., 2000. Belgium: language, ethnicity and nationality. *Parliamentary Affairs*, 53(1), pp. 114–134.
Rokkan, S., 1999. *State formation, nation-building, and mass politics in Europe*. Oxford: Oxford University Press.

Scott, J.C., 1998. *Seeing like a state: how certain schemes to improve the human condition have failed.* New Haven, CT: Yale University Press.

Servais, J. and Mechelynck, E., 1933. *Les codes et les lois spéciales les plus usuelles en vigueur en Belgique: avec des notes de concordance et de jurisprudence utiles à l'interprétation des textes.* Brussels: Éditions Émile Bruylant.

Simon, P., 1999. Nationality and origins in French statistics. Ambiguous categories. *Population: An English Selection*, 11, pp. 193–219.

Torpey, J., 2000. *The invention of the passport: surveillance, citizenship and the state.* Cambridge: Cambridge University Press.

Vandenbussche, W., 2007. Wij willen Willem weer: het Verenigd Koninkrijk der Nederlanden als breekijzer voor een nieuwe 19de-eeuwse taalgeschiedenis. In: W. Vandenbussche and G. Janssens, eds. 2007. *Taalpolitiek, taalplanning en taalgebruik in het 'Verenigd Koninkrijk der Nederlanden' (1814–1830).* Ghent: Koninklijke Academie voor Nederlandse Taal- en Letterkunde. pp. 129–135.

Vandermoere, F. and Vanderstraeten, R., 2012. Disciplinary networks and bounding: scientific communication between science and technology studies and the history of science. *Minerva*, 50(4), pp. 451–470.

Vanderstraeten, R. and Vandermoere, F., 2015. Disciplined by the discipline: a social-epistemic fingerprint of the history of science. *Science in Context*, 28(2), pp. 195–214.

Verdoodt, A., 1983. *Bibliographie sur le problème linguistique Belge.* Québec: Centre International de Recherche sur le Bilinguisme.

Vogl, U. and Hüning, M., 2010. One nation, one language? The case of Belgium. *Dutch Crossing. A Journal of Low Countries Studies*, 34(3), pp. 228–247.

Wade, P., 2014. Race, ethnicity and technologies of belonging. *Science, Technology, & Human Values*, (39)4, pp. 587–596.

Wils, L., 2009. *Van de Belgische naar de Vlaamse natie: Een geschiedenis van de Vlaamse beweging.* Leuven: Acco.

9 'Make me happy and I shall again be virtuous'

Science fiction and the utopian surplus of science

Tom Moylan

Cognitive estrangement

In my contribution to this multidisciplinary examination of the social function and impact of collective imaginaries of the future, I will consider how 'sciences, and scientific imagination … come to shape social imagination', and vice versa.[1] As I do so, I will bring the critical perspectives and methodologies of literary and utopian studies to bear on my discussion of the ways in which science fiction – as an aesthetic form deeply engaged with science – can produce both a critique of the present and a utopian anticipation of an improved life for humans and nature. I will begin by taking a close look at Mary Shelley's *Frankenstein* (one of the earliest works of modern science fiction) to examine what this iconic work has to say about science, society and the role of the scientist. Then I will turn to the science fiction of our own era in order to demonstrate that this 'privileged' literature of modernity (as Fredric Jameson (2005) has called it) has continued the diagnostic and prognostic work of which this genre is capable. Again, I will look at a single exemplary text, in this case Marge Piercy's *Woman on the Edge of Time* (1976), a writer and a novel with strong affinities to Shelley and her work.

Mary Shelley's 1818 novel is often cited as the *originary* text of modern science fiction, but of course its place in literary history is more complex than that. For, while *Frankenstein* is a milestone in the development of the genre, it arose out of the intertextuality of earlier writing created within the historical dynamics of the Western world. Some scholars locate predecessors as far back as the imaginary voyages and scientific poetry of Lucian of Samosata in first century Syria, but more scholars pick up the story with Johannes Kepler's *Somnium* in the early seventeenth century.[2] From there, we can follow the evolution of this literary form as it moves away from fantastic imaginaries into the modern discourse of speculative reason. By the eighteenth century, tales of voyages to the moon occasioned commentaries on the worlds of the authors, as they considered scientific and technological matters as well as societal dynamics and problems.[3] In the mid- to late eighteenth century, another narrative channel was opened by the Irish writer Samuel Madden and the French writer Louis Sébastian Mercier when each introduced the trope of the *future* as a literary device that provided a distanced temporal (indeed historical) perspective from which the complexities of the present could be addressed.[4] More fundamentally, the tradition of literary utopias, from Thomas More's eponymous work in 1516 through to the dystopian satires of the likes of Jonathan Swift and Voltaire provided a critical backdrop for this new writing.[5] By the nineteenth century, therefore, authors influenced by Shelley continue to create narrative fictions shaped by this aesthetic, in which an anticipatory realism imbued with a scientific sensibility was capable of transcending the conceptual and discursive limits of the present.

Before turning to the text of *Frankenstein*, I want to review the basic characteristics of this genre. While science fiction (or sf) has always been highly entertaining, at a deeper level it also carries a particular capacity for interrogating modern society. In this regard, the science fiction imaginary should not be equated with technocratic expressions of futurism or with popular culture predictions of scientific or technological trends. Rather, sf offers an imaginative approach that refuses to regard the present as static or provincial, seeing it instead as ripe with possibilities, good or ill, not yet realized.[6]

In this creative process, science fiction generates thought experiments that work by way of extrapolation or analogy, and often both. That is, a science fiction text can explore the immediate possibilities and dangers of social and scientific conditions by spinning them slightly forward – here we can think of representational narratives of biological engineering or space travel in a future in which our present is its past.[7] But a work of science fiction can also build an entirely other world that is far distant from that of its author and yet one that offers an analogous figuration of it: here, the future histories of Isaac Asimov or the television series *Battlestar Galactica* come to mind. Whether a given text explores the very near future, as in *Frankenstein* or the cyberpunk novels of William Gibson, or a far distant spatio-temporal reality as portrayed by H.G. Wells or Iain M. Banks, science fiction almost always creates an alternative reality that resonates with the author's present. This formal and conceptual operation has been named by Darko Suvin (1979) (drawing on the aesthetics of Bertolt Brecht) as *cognitive estrangement*, that is, an exercise of speculative reason that (re)visions the known world as it is refracted through an imaginary space-time. By distancing readers from that world so that they can take a second and hopefully more probing look, sf enables a critical and creative engagement with it.[8]

Underlying this imaginative mechanism, sf is driven by a utopian imaginary that is unafraid of considering what can get worse (the dystopian mode) or what could be done better (the eutopian mode).[9] Whether in a singular composition based on probing extrapolations (for example, by way of robotics, computer technology or apocalyptic devastation brought on by nuclear holocaust, population explosion or rapid climate change) or in more complex symphonies of social transformation (as in the literary utopia or the alternative history novel), science fiction to a greater or lesser extent operates along the continuum of dream and nightmare carried within the unfolding history of the modern world. I ask you, therefore, to bear these generic characteristics in mind as I turn to Shelley's exploration of the promises of science and the responsibility of the scientist.

Mary Shelley's warning

In the summer of 1816, while on holiday in Switzerland, Mary Shelley, her husband Percy Bysshe Shelley, the poet Byron and their friend Polidori agreed to a ghost story competition around their fire. The result Mary delivered the next day was the first draft of the epical *Frankenstein*, which would move the cultural imagination, informed by the critical attitudes of romanticism, away from the literary strains of the Gothic tale into an as yet unnamed form. While not all sf focuses on science, Shelley's novel in particular examines the relationship of science to the contradictions and injustices of a rapidly changing society.

Without going into an extensive biographical review, let me recall that Mary, at nineteen, had been in a relationship with Shelley for two years before their travels in Switzerland. The daughter of the anarchist William Godwin and the feminist Mary Wollstonecraft, her sense of the world was already shaped by the radical values that infused early nineteenth-century Europe. With Shelley and their friends, she was of the generation of intellectuals and artists who embraced the spirit of the French Revolution: liberty (intellectual and sexual), equality (economic and social), and fraternity (with the emergent working class, women and those who were enslaved or colonized). Living out their beliefs in bohemian style and affiliated with political campaigns against slavery and for Chartism, Irish freedom and women's rights, they valued critical reason over received belief and each in their own way sought to create artistic expressions that were true to the culture of romanticism.

This radical milieu informed Shelley's writing in content and form. Growing up with her mother's commitment to feminism, she was well aware of the oppressed position of women in her society (experiencing it herself in her own career as a writer), and this gendered conflict of subjugation and struggle provides a powerful subtext in her novel. She was also politically and aesthetically influenced by her father. In particular, she carried the spirit of his realist novel *Caleb Williams* – in which a common man is destroyed by his bourgeois master – into her new speculative mode of writing, even as she also drew on the canonical narratives of Prometheus and his break from Zeus, Satan and his break from the Christian God, and the tortured trajectory of Faust. More broadly, Enlightenment and romantic values such as the goodness of natural man and the liberatory power of education inform her story of a new sentient being produced by experimental science. Significantly, therefore, she draws on the scientific research that she and her circle were discussing (such as Humphrey Davies's chemistry, Erasmus Darwin's botany, and Luigi Galvani's physics) to ensure that hers would be a tale not of supernatural power but of scientific invention. On this tapestry of ideas and values, practices and anticipations, Shelley lays out an enduring account of the promise and failure of science and the scientist, and of the consequent alienated existence experienced by the new subjects of history.

I'll begin with the most familiar passage from *Frankenstein*, one that metonymically expresses the project of the entire novel. This is the moment when

Victor Frankenstein brings his creature to life, the moment when the promise of science begins to segue into its failure:

> It was on a dreary night of November, that I beheld the accomplishment of my toils. With an anxiety that almost amounted to agony, I collected the instruments of life around me, that I might infuse a spark of being into the lifeless thing that lay at my feet. It was already one in the morning; the rain pattered dismally against the panes, and my candle was nearly burnt out, when, by the glimmer of the half-extinguished light, I saw the dull yellow eye of the creature open; it breathed hard, and a convulsive motion agitated its limbs.
>
> (Shelley, 1974, p. 52)

In terms of style, the tone of this passage would readily resonate with contemporary readers who loved their Gothic novels: the dreary night, dismal rain, flickering candle and dull yellow eye of a creature made up of body parts harvested from cadavers all evoke this popular mode. But, in a brilliant exercise of what Raffaella Baccolini (2000) identifies elsewhere as the writerly process of 'genre blurring', in this passage, and then throughout the novel, Shelley conjoins what Kathleen Eull has described as 'the descriptive heft of a Gothic scene set and the terseness of a scientist's journal entry' by way of an artistic stitching 'as carefully crafted as the sutures of Victor Frankenstein himself'.[10] In doing so, Shelley shifts generic gears into the new literary mode of realism in which the modern world is approached without an aura of myth or mystery. However, she then makes a pathbreaking second shift as she transforms that fashionable realism into a realistically speculative imaginary that enables her to explore her society at an existential and systemic level.

Considering the novel in terms of its extrapolative register, Shelley creates a narrative in which the science of her day leads not only to the creation of life, but to a superior form of human life. At this level, the story evinces its existential quality as it comes to us through the words of the scientist himself. The son of bourgeois parents who provided him with a comfortable childhood in Switzerland, Frankenstein matriculates at the University of Ingolstadt and enters into a study of the medieval sciences of Cornelius Agrippa, Paracelsus and Albertus Magnus. But, in a telling cultural and political development, he is challenged by the male authority figures in his life – namely, his father and his professors – to set aside the older paradigm and embrace what his father calls the 'modern system of science', especially since it can yield results which are 'real and practical' (Shelley, 1974, p. 33). Yet, even as he moves into his studies of chemistry and electricity, he carries with him the fascination with the nature of life that was ignited by his earlier reading. He eventually succeeds in discovering the science and technology of, as he puts it, 'bestowing animation upon lifeless matter' (see Shelley, 1974, p. 46ff.). Thus, his alchemical dream segues into scientific achievement. Now obsessed by an almost godlike power, he sets out with prideful determination to create a human being:

> A new species would bless me as its creator and source; many happy and excellent creatures would owe their being to me.... Pursuing these reflections, I thought that if I could bestow animation upon lifeless matter, I might in the process of time ... renew life where death had apparently devoted the body to corruption.
>
> (Shelley, 1974, p. 49)

Working intensively, Frankenstein assembles body parts gathered from the 'unhallowed damp of the grave' in order to produce a human being that is brought to vitality by the new science of electricity. Such is his achievement on that dreary night. But despite the horrific *tone* of Frankenstein's words, it is crucial to note that his creature is a product of science, not magic. In keeping with the imperative for 'improvement' common to enlightened thought of his time, the scientist has made what he identifies as a 'human being in perfection' (Shelley, 1974, p. 51). Tall in stature and endowed with great strength and endurance, with acute hearing and sight and a powerful intelligence that endows it with the capacity for rapid learning and complex conceptualization, the creature may be radically different from humanity as Frankenstein and others of his time may know it, but he is not inherently monstrous. Indeed, what lies before Frankenstein on that dark night is, in this initial moment at least, a product of the dream, not the nightmare, of modernity.

There is, therefore, no *immediate* reason for Frankenstein to reject his creation; and yet, as Zeus and God did with Prometheus and Satan, this young scientist will fail to accept the life he has created. His abandonment, however, has more to do with the historical and material conditions of his time than with mythic or epic struggle: for the existential clash Frankenstein faces is between his enlightened impulse to discover the secret of life and contribute to the improvement of the human race and his affiliation with the socio-economic agenda of his own class. As he beholds the results of his work, he begins to realize that this being would mark the negation of society as he knows it. For this is not the docile subject that could have been generated by the 'real and practical' science advocated by his father, but rather a radically new being capable of claiming his own agency and place in the world. At this breakthrough moment, in which an emancipatory science could have yielded a transformative result that broke beyond the economic and political order of things, Frankenstein sees 'the beauty of the dream' vanish (Shelley, 1974, p. 53). Rather than accept the creature in his own right, he acts from the perspective of his privileged gaze and interpellates the creature as a 'miserable monster' (Shelley, 1974, p. 53). He then averts his eyes and turns his back, condemning his creation to a life of alienation and abjection. And so it is that, early in the nineteenth century, Mary Shelley captures the way in which science can be compromised by the actions of a scientist who is unable to look beyond the normative respectability required by what Pierre Bourdieu (1972) would identify as his 'bourgeois habitus'.

While Frankenstein's account of events delivers this extrapolative narrative of scientific practice, in the analogical register Shelley connects this story to the historical failure of the social system of her day to produce a more just and egalitarian society because it cannot break out of the iron cage of its own self-interest. At the core of the novel, therefore, the narrative of scientific responsibility entwines with the evocation of the creature as the bad subject who represents all the emergent human beings produced and exploited by the world system of industrial capitalism, state power and white male privilege: namely, the new proletariat, women, chattel slaves and indigenous populations who have been colonized or genocidally eradicated – all of whom can be recognized under the general category of 'subaltern', a term I take from postcolonial studies that identifies those who are socially, politically and geographically outside of a hegemonic power structure.[11] As this narrative unwinds, what Ernst Bloch (1986) identifies as the 'utopian surplus' that is inherent in any social system becomes evident in Shelley's text through the words of the creature; for, while Frankenstein's reaction reveals the dystopian nature of Shelley's own world, the superior but alienated creature speaks from a standpoint that faces a better horizon.

This systemic exposé begins to unfold when the creature returns from his initial exile to confront Frankenstein with his demand for recognition and justice. As with most utopian expressions, which begin in negation and then unfold its positive alternative, he tells his creator what he has learned about history and present-day reality. Having mastered not only the practices of everyday life but also language and literacy through his clandestine observations of the De Lacey family, he draws on his furtive reading of their books in order to comprehend history, theology and ethics.[12] He therefore conjoins his growing sense of self-identity with a knowledge of not only the condition of his own alienation but also that of others without power. In just a few words, he describes the socio-economic relations and processes of subject construction that have rendered him an outcast:

> I heard of the division of property, of immense wealth and squalid poverty; of rank, descent, and noble blood.... I learned that the possessions most esteemed by your fellow-creatures were, high and unsullied descent united with riches. A man might be respected with only one of these acquisitions; but without either he was considered, except in very rare instances, as a vagabond and a slave, doomed to waste his powers for the profit of the chosen few.
>
> (Shelley, 1974, p. 115)

In a later confrontation with Frankenstein, still holding to his newly learned principles of justice, the creature again attempts to negotiate in good faith:

> Life, although it may only be an accumulation of anguish, is dear to me, and I will defend it. Remember, thou has made me more powerful than thyself; my height is superior to thine; my joints more supple. But I will not be tempted to set myself in opposition to thee. I am thy creature, and I will be

> even mild and docile to my natural Lord and King if thou wilt also perform thy part, that which thou owest me.
>
> (Shelley, 1974, p. 95)

Here, on the frozen Arctic waste, the subaltern speaks back to his creator. The new subject of history calls for respect and equity. Having thrown off his imposed identity as a monster, he pleads for justice: 'Every where I see bliss, from which I alone am irrevocably excluded. I was benevolent and good; misery made me a fiend. Make me happy, and I shall again be virtuous' (Shelley, 1974, p. 95). In this moment, however, the creature conjoins his earlier critique of existing society with a new demand as he asks Frankenstein to create a woman to be his companion, so Adam may have an Eve. From this added claim arises the valorized possibility of a new species and a new society, as we see figured in the creature's vision of the pastoral life he and his bride would build in the 'vast wilds of South America' (Shelley, 1974, p. 142). In language that echoes the familiar discourse of the utopia, he describes this *novum* of a 'peaceful and human' existence free of misery, wherein these new humans would live on acorns and berries and sleep on beds of dried leaves under a warming sun (Shelley, 1974, p. 142).

Momentarily filled with compassion, Frankenstein agrees to the creature's request, but when he returns to the security and comfort of his home in Geneva he redoubles his rejection as he decides not to carry out this second creation. Now fully asserting his own class affiliation, he sees himself protecting 'existing generations' and 'future ages' from what threatens to become the collectivity of 'a race of devils' (Shelley, 1974, p. 163). When the creature learns of this final denial, he reverts to his threat to destroy Frankenstein and those he loves, not out of an original sin of monstrosity but out of the injustice of his immiseration: 'if I cannot inspire love, I will cause fear; and chiefly towards you my arch-enemy, because my creator, do I swear inextinguishable hatred. Have a care: I will work at your destruction, nor finish until I desolate your heart' (Shelley, 1974, p. 141). Thus, the subaltern strikes back, and Shelley's narrative unwinds in a spiral of personal tragedy and social destruction. The creature wreaks his vengeance against Frankenstein's loved ones, and the scientist again pursues him onto the Arctic ice, only to die by the hands of his creation. In a final exhortation to Walton, the explorer who delivers the framing narrative at the end of the novel, the creature recalls his origins in the service of an apparently reasonable science and condemns his own suffering as well as that which he imposed on others. Utterly alone, the subaltern, the rejected creature, turns from Walton and wanders away. In the last words of the novel, he is 'lost in darkness and distance' (Shelley, 1974, p. 221).

Marge Piercy's vision

Moving on from *Frankenstein*, I now want to look at the science fiction of our own era as it continues to explore the relationship of the scientific and the social.

Certainly, I could follow the dystopian fiction that focuses on the mad or evil scientist, as we can trace it from Shelley through work such as H.G. Wells's *Island of Doctor Moreau* down to Stanley Kubrick's film *Dr. Strangelove* (1964). For today, however, I will focus on the *eutopian* tradition that reaches through the dystopian present to the horizon of better alternatives – doing so in a literary stream that runs from nineteenth-century novels such as William Morris's *News from Nowhere* and Charlotte Perkins Gilman's *Herland* to the utopian science fiction written since the Second World War. In these later works, the unfulfilled utopian surplus of modernity embedded in the Enlightenment, in the French Revolution and in scientific and industrial progress was recaptured by changing historical and political conditions. For in the political movements from the 1940s onwards, unhampered state power, the mechanisms of late capitalism and the institutional practices of racism, sexism and other forms of human and natural oppression were challenged by oppositional movements that were often led by the subaltern subjects who emerged out of the prevailing system of production and exploitation.[13]

Crucially at this moment, science itself was at the centre of a contest between a practice deeply entrenched in the dominant political, economic and, of course, military systems and one which opposed that co-optation in the name of serving humanity and nature.

One of the most widely read and well-regarded works in this new utopian movement is Marge Piercy's *Woman on the Edge of Time*. Whereas in *Frankenstein* the thought experiment regarding the creation of human life ends in disaster, Piercy's novel sustains Shelley's attention to the flawed imbrication of science and society but then takes a further step and creates a fully realized eutopian alternative. Published in 1976, at the height of the New Left/feminist/ecological movements, *Woman on the Edge* challenges the institutional power of science as it serves the interests of the existing social system but also reclaims the utopian promise that it can deliver, as science studies scholar Sandra Harding puts it, 'social progress to all the world citizens rather than … to an elite few' (Harding, 2008, p. 2, p. 19). Pulling this systemic challenge through her text is Piercy's existential tale of one person who, unlike the creature, moves from a position of victimhood to one of radical utopian agency.

Like Shelley's, Piercy's writing grew out of her own journey as she discovered her artistic and political vocations in times of intense social and political change. Active in New Left and socialist feminist politics of the 1960s, she went on in the 1970s to work with the new ecological movement and with the Women's Health Project and the Mental Patients Liberation Front. As a writer of poetry, realist novels and science fiction, she also creatively engaged with the personal and the political conditions and struggles of this period. Drawing these two strands together, she was part of the movement with other science fiction writers such as Ursula K. Le Guin and Samuel R. Delany that revived, indeed refunctioned, the utopian novel so that it shifted from an emphasis on utopia as a set blueprint to one that focused on the process of moving towards and living in the better society. Given this shift, it was no surprise that these 'critical utopias' paid close attention to the transformative politics required to move from the bad old world to the new.[14] Further to our theme, it is especially important to note that each of these works addresses the epical tension between Faustian failure and Promethean rebellion as they examine the competing instrumental and emancipatory roles of science.

Unlike *Frankenstein*, however, wherein utopian hope is reduced to the creature's declarations of desire and his sketch of a better life in South America, in *Woman on the Edge* Piercy, writing in a time of more fully developed political alternatives, not only relates the biography of her protagonist, Connie Ramos, but also details her ethnography of a utopian alternative.[15] While in her dystopian present she focuses on the role played by neuropsychology within the penal-scientific complex of the US social welfare and mental health system, in her future society she explores how 'appropriate science and technology' (to use a familiar phrase from the 1970s) can help to realize its full utopian potential, especially with the help of an ecologically attuned practice of bioengineering.[16]

Like Shelley's creature, Connie is both a character in her own right and a figure of all subalterns. She is a Mexican immigrant in New York City, poor and unemployed, and a single parent whose daughter has been taken from her by a welfare bureaucracy that convicted her of cruelty when in fact it was poverty and exploitation that did the damage. With her daughter placed in the care of a middle-class suburban family, she is committed to Rockover State Psychiatric Hospital. There, Frankenstein's rejection is reprised by the psychiatric experts who deny and destroy the dignity and integrity of her person. After Connie resists the standard interventions of sedative drugs and electroshock treatment, she is selected (along with others such as an older black woman and a young gay man) for an experiment in the control of those the system labels as 'socially violent'. She soon learns that researchers from the New York Neuropsychiatric Institute have developed a brain implant with a computerized feedback mechanism that enables doctors to intervene when the subject acts in a manner which they, as the guardians of mainstream society, deem inappropriate. This biodisciplinary project of containing and controlling such 'miserable monsters' becomes the primary plot device for Piercy's critique of the way science has been co-opted: here, social justice is trumped by social control, a profitable control because implantation can minimize the costs of surveillance and supervision. In a terse description that echoes Frankenstein's ideological desire for a docile subject, an inmate identifies the implant procedure as one that will 'turn us into machines so we obey them' (Piercy, 1976, p. 192).[17]

Piercy's utopian negation of this braided relationship of science and power begins when Connie is contacted by a visitor from the future. Lucíente (who is a utopian version of Connie) has travelled through time from her village of Mattapoisett to enlist her in the historical struggle against what she calls 'the Age of Greed and Waste' (Piercy, 1976, p. 49). In the typical trajectory that has informed utopian narrative since Thomas More, Connie is introduced to all the components of this society by Lucíente and her fellow villagers. She learns that this modest-looking society is comprised of decentralized villages governed by a participatory form of democracy, based in an anarcho-communist political economy, sustained by a biologically based high technology, and committed to the eco-socialist principles of human liberation and ecological flourishing. In this context, each citizen is enabled to develop holistically while contributing each in their own way to the community's optimal functioning. Production practices, decision-making, personal relations, sexuality, parenting, education, punishment, spirituality and language are all informed by the ethos of this self-reflexive and self-correcting utopian totality.

Here, I want to highlight the role played by appropriate science and technology in Mattapoisett. Lucíente (who is a plant geneticist) describes how their collectively decided and practiced science engages with the values of their social system: procedures and decisions are monitored, reviewed and debated in light of their impact on the overall unfolding of their utopian life. Computer technology provides automation and feedback mechanisms that enhance rather than desiccate everyday life and production. A highly developed energy technology employs renewable sources that include solar, methane, wind power, water power and tidal power. And, against the grain of current environmentalist thinking, a monitored practice of genetic engineering enables the utopians to privilege diversity over monoculture, to ensure a flourishing natural environment and to maximize human health and social complexity (by way of genetic interventions that range from the repair of vision and the reversal of cancer to the minimalist design of embryos and the enablement of agency in other nonhuman sentient animals). In addition, advanced medical procedures (such as microsurgery and, controversially, a system of extrauterine production that achieves what feminist theorist Shulamith Firestone described as the final step of equality for women) work compatibly with folk medicine (including herbal and mental healing) to ensure that all of nature, including humanity, is nurtured and enabled.[18] Indeed, this utopian science–social complex reproduces new human beings in a way that stands in stark contrast to the fate of the creature: for the combination of genetic design, extrauterine reproduction and advanced medical techniques with a loving and therapeutic cultural environment that involves parents, aunts and uncles, and the entire village results in children who are wanted and cared for and involved from an early age in community life.

While these scientific and technological practices would be recognized by a 1970s reader as extrapolations of existing tendencies, the driving narrative device of time travel is, of course, an analogical figure, one evoking the poetics of history, not science. Piercy's science-fictional premise of trans-temporal communication makes possible her political meditation on the necessity for individual and collective activism in the present for the future attainment of a better society. This political lesson becomes evident after Connie has begun to understand and appreciate the utopian society. During one of the later visits, Lucíente explains to Connie that their future is not guaranteed. Typical of the critical utopian tendency to portray the better society not as static blueprint but as process, the reality of Mattapoisett is still dependent on the work of history: that is, it can only exist, or come to exist, if people of Connie's time and others act responsibly and effectively. Consequently, Connie is invited to join in solidarity with the utopian community by making her contribution as a political agent. As Lucíente explains: 'We must fight to come to exist, to remain in existence, to be the future that happens. That's why we reached you' (Piercy, 1976, p. 190). At this point, the utopian surplus of existential and systemic hope anticipated by, yet denied to, Frankenstein's creation is not only revived but carried to a radically successful outcome as Connie responds to this challenge with a fulfilled sense of self that was unavailable to Shelley's abject creature.

The novel ends with a series of metonymic events that serve the larger purpose of Piercy's parable, for Connie's mission is to return to the psychiatric hospital and destroy the experiment that is a key component in the evolution of a more efficient regime of what Herbert Marcuse (1964) later would call the 'totally administered society' and Michel Foucault would term 'biopower'.[19] In a dénouement that some readers find disturbing, Connie assassinates the four principal investigators and by implication stops the project, both in itself and as a sign of the scientifically imbricated power relations of the entire society. As a result, she is incarcerated for life in the state hospital, a confinement registered by the case notes that fill the closing pages of the book. Yet, while Connie's body is trapped in a 'darkness and despair' that recalls the dystopian fate of the creature, the utopian seed that was at the heart of Shelley's novel has grown in this critical utopia into a radically other life that produces a liberated subjectivity capable of speaking in a fully utopian voice.

A promise hard to keep

I hope that my readings of these two major works of science fiction have exemplified the ways that this genre draws from and speaks back to science and technology. More pointedly, I hope I've shown how this interrelationship infuses creative works that address the compromises and co-optations of science as well as its not yet fulfilled utopian potential.

There are, of course, many other writers who engage with this question. For example, in Ursula K. Le Guin's *The Dispossessed* (1974) a scientist from a utopian society (who is an analogue of Einstein or Sakharov) challenges the self-interest of the scientific establishment in the process of arriving at a general temporal theory which enables instantaneous communication across the galaxy; in doing so, Shevek expedites the formation of a galactic federation based on the principles of what Jürgen Habermas calls communicative reason. And, in Kim Stanley Robinson's *Science in the Capital Trilogy*, the apocalyptic destruction of the Gulf Stream caused by rapid climate change is reversed by the systemic intervention of a scientifically informed politics (led by an enlightened US president and an activist in the National Science Foundation) and the 'big science' of geo-engineering (or 'terraforming').[20]

For today, however, I'll leave you with words of critique and hope from Robinson's recent novel *Galileo's Dream* (a time travel story wherein Galileo is transported to the future by people who regard him as the First Scientist so as to help prevent another round of global destruction abetted by compromised scientific practice). As Galileo himself experiences his own cognitive estrangement brought about by travelling to the future, he reflects critically upon history and summarizes the conflict that I have been addressing throughout this talk:

> Science was broke and so it got bought. Science was scared and so did what it was told. It designed the gun and gave the gun to power, and power then held the gun to science's head and told it to make some more. How smart was that? Now science is in the position of having to invent a [new weapon], and then start the whole process over. It's not clear it can work.... If we can shift that power.... That's the if. If we can shift history into a new channel, and avoid the nightmare centuries. If we can keep the promise of science, a promise hard to keep.
>
> (Robinson, 2009, pp. 575–576)

Yes, this is a promise hard to keep. But, if we are ever to reach beyond the political, economic and ecological crises in which we are all embroiled, it is what must be done – existentially and systemically, personally and politically.

Notes

1 I want to thank Kathleen Eull and Katie Moylan for their incisive comments on earlier drafts of this chapter.

2 See Lucian of Samosata, *True History* (*c.* 150 AD; English translation 1634) and Johannes Kepler, *Somnium* (1608).

3 See for example the classic study by Nicolson (1960).

4 Specifically: Samuel Madden, *Memoirs of the Twentieth Century* (1733) and Louis Sébastian Mercier, *L'an 2440* (1770).

5 For good introductions to the relationship between utopianism and utopian literature, see Sargent (2010), and Claeys (2010).

6 For good introductions to science fiction and science fiction theory and criticism, see Bould, Butler, Roberts and Vint (2009), and Clute and Nicholls (1993).

7 For a useful discussion on the thought experiment in science fiction, see Le Guin's 'Introduction to the Left Hand of Darkness' (1979).

8 Working from this capacity for sf to facilitate a cognitive and critical assessment of the author's and reader's world, Freedman (2000) further demonstrates the depth of the relationship between science fiction and contemporary critical theory by arguing that each offers a version of the other.

9 It is important to note that the utopian process requires that the move to a better social reality should be a *totalizing* one, that is, a systemic transformation of an entire society. Piecemeal reform is only a significant component of this process if it is a considered step in that larger transformative movement.

10 Kathleen Eull, email communication, 31 January 2015.

11 On the subaltern, see the work of subaltern studies and especially the key essay by Spivak, 'Can the Subaltern Speak?' (1988).

12 Specifically: Volney's *Ruins of Empire*, Milton's *Paradise Lost*, Plutarch's *Lives* and Goethe's *Sorrows of Werther* (see Shelley, 1974, p. 123ff.).

13 There is also a science-fictional line of writing that focuses on scientific development in a positive, even enthusiastic, manner. Such work, which in the critical terminology is called 'Hard SF', tends to focus on the practices and not the problems of science.

Common themes have been space travel and engineering as well as cybernetics and genetics; but the most enduring has been that of military development (which one can trace from the 'future war' novels of the late nineteenth century to narrative and visual portrayals of space war by writers such as Gerry Pournelle and Gregory Benford and in films such as the *Terminator* series (1984)). This is not the place to download a bibliography, but two valuable studies of the military stream are Clarke's *Voices Prophesying War* (1996) and Franklin's *War Stars* (1988).

14 As I put it in my initial study of the critical utopia (by Piercy (1976), Le Guin (1974) and Delany (1976), but also Joanna Russ (1975), Ernest Callenbach (1975) and others),

> utopian writing in the 1970s was saved by its own destruction and transformation into the 'critical utopia'. 'Critical' in the Enlightenment sense of critique – that is expressions of oppositional thought, unveiling, debunking, of both the genre itself and the historical situation. As well as 'critical' in the nuclear sense of the critical mass required to make the necessary explosive reaction.... A central concern in the critical utopia is the awareness of the utopian tradition, so that these texts reject utopia as blueprint while preserving it as dream. Furthermore, the novels dwell on the conflict between the originary world and the utopian society opposed to it so that the process of social change is more directly articulated. Finally, the novels focus on the continuing presence of difference and imperfection within utopian society itself and thus render more recognizable and dynamic alternatives.
>
> (Moylan, 1986, pp. 10–11)

15 As readers begin Piercy's narrative, they may first comprehend it in the terms of the modern realist exposé novel (as in works such as Ken Kesey's *One Who Flew Over the Cuckoo's Nest*, 1963); but when the speculative science premise of the brain implant enters the story, readers are required to switch into the literary protocol of science fiction. Left strictly within the experiment narrative, Connie's story would easily take its place alongside that body of work uncovering the conditions of life in post-war America that Kingsley Amis (1960) termed the 'new maps of hell'. But, as with the formal shifts in Shelley's novel, Piercy does not stop at describing a map of hell. Writing in the 1970s moment of engaged hope, she breaks through this science-fictional portrayal of dystopian darkness and creates a eutopian space that changes the overall impact of her novel.

16 On 'appropriate science and technology', see Sismondo (2003).

17 The resonance here with Foucault's analysis of the disciplinary society is quite strong; however, I have no evidence that Piercy was reading Foucault at this time. While his *Surveiller et punir: Naissance de la Prison* was published in 1975, the English translation only became available in 1977.

18 Echoing the spirit of Firestone's argument (1970), Lucíente explains their society's decision to reproduce by way of extrauterine reproduction:

> it was part of women's long revolution. When we were breaking all the old hierarchies. Finally, there was that one thing we had to give up too, the only power we ever had, in return for no more power for anyone. The original production: the power to give birth. 'Cause as long as we were biologically enchained, we'd never be equal. And males would never be humanized to be loving and tender. To break the nuclear bonding.
>
> (Piercy, 1976, p. 98)

19 On biopower, see Foucault (1976), and Hardt and Negri (2000). On the relationship between systemic power and science and technology, see Aronowitz (1988).

20 Cultural critics such as Sherryl Vint (2012) have linked the problematic of science fiction with that of science studies: see her entry in Bould *et al.* (2009). And, in a recent essay, Vint investigates the potential of what Bruno Latour has called an

'amodern' science that, especially as informed by feminist and non-Western theory and practice, would refuse 'an artificial separation between science and society, one that recognizes that the values by which we organize our social and material lives will also shape the kind of science we have and the futures it can help us produce' (Vint, 2012, p. 46).

References

Amis, K., 1960. *New maps of hell: a survey of science fiction.* New York, NY: Harcourt Brace.

Aronowitz, S., 1988. *Science as power: discourse and ideology in modern society.* Minneapolis, MN: University of Minnesota Press.

Baccolini, R., 2000. Gender and genre in the feminist critical dystopias of Katherine Burdekin, Margaret Atwood, and Octavia Butler. In: M. Barr, ed. 2000. *Future females, the next generation: new voices and velocities in feminist science fiction.* Boston: Rowman and Littlefield. pp. 13–34.

Bloch, E., 1986. *The principle of hope.* 3 vols. (trans. N. Place, S. Plaice and P. Knight). Cambridge, MA: MIT Press.

Bourdieu, P., 1972. *Outline of a theory of practice* (trans. R. Nice). Cambridge and London: Cambridge University Press.

Bould, M., Butler, A.M., Roberts, A. and Vint, S., eds., 2009. *The Routledge companion to science fiction.* London and New York, NY: Routledge.

Callenbach, E., 1975. *Ecotopia.* New York, NY: Bantam.

Claeys, G., ed., 2010. *The Cambridge companion to utopian literature.* Cambridge and New York, NY: Cambridge University Press.

Clarke, I.F., 1996. *Voices prophesying war: future wars 1763–3749.* New York, NY, and London: Oxford University Press.

Clute, J. and Nicholls, P., 1993. *The encyclopedia of science fiction.* London: Orbit.

Delany, S.R., 1976. *Triton: an ambiguous heterotopia.* New York, NY: Bantam.

Dr. Strangelove or: how I learned to stop worrying and love the bomb, 1964. [Film] Directed by Stanley Kubrick. USA: Columbia Pictures.

Firestone, S., 1970. *The dialectic of sex.* New York, NY: Bantam.

Foucault, M., 1975. *Surveiller et punir: naissance de la prison.* Paris: Gallimard. English edition: *Discipline and punish: the birth of the prison* (trans. A.Sheridan, 1997). New York, NY: Pantheon.

Foucault, M., 1976. *The history of sexuality. Vol. 1: the will to knowledge.* London: Penguin.

Franklin, H.B., 1988. *War stars: the superweapon in the American imagination.* New York, NY, and London: Oxford University Press.

Freedman, C., 2000. *Critical theory and science fiction.* Hanover, CT: Wesleyan University Press.

Harding, S., 2008. *Sciences from below: feminisms, postcolonialities, and modernities.* Durham, NC, and London: Duke University Press.

Hardt, M. and Negri, A., 2000. *Empire.* London and Cambridge, MA: Harvard University Press.

Jameson, F., 2005. *Archaeologies of the future: the desire called utopia and other science fictions.* London: Verso.

Kesey, K., 1963. *One flew over the cuckoo's nest.* New York, NY: Signet.

Le Guin, U.K., 1974. *The dispossessed: an ambiguous utopia*. New York, NY: Harper & Row.

Le Guin, U.K., 1979. Introduction to the left hand of darkness. In: U.K. Le Guin, 1979. *The languages of the night: essays on fantasy and science fiction*. New York, NY: Perigee. pp. 155–161.

Marcuse, H., 1964. *One dimensional man*. Boston, MA: Beacon.

Moylan, T., 1986. *Demand the impossible: science fiction and the utopian imagination*. New York, NY, and London: Methuen.

Nicolson, M.H., 1960. *Voyages to the moon*. New York, NY: Macmillan.

Piercy, M., 1976. *Woman on the edge of time*. New York, NY: Knopf.

Robinson, K.S., 2004. *Forty signs of rain*. New York, NY: Bantam.

Robinson, K.S., 2005. *Fifty degrees below*. New York, NY: Bantam.

Robinson, K.S., 2007. *Sixty days and counting*. New York, NY: Bantam.

Robinson, K.S., 2009. *Galileo's dream*. London: Harper.

Russ, J., 1975. *The female man*. New York, NY: Bantam.

Sargent, L.T., 2010. *Utopianism: a very short introduction*. Oxford: Oxford University Press.

Shelley, M., 1974. *Frankenstein*. New York, NY: Bobbs-Merrill.

Sismondo, S., 2003. *An introduction to science and technology studies*. New York, NY: Wiley.

Spivak, G., 1988. Can the subaltern speak? In: C. Nelson and L. Grossberg, eds. 1988. *Marxism and the interpretation of culture*. Urbana and Chicago, IL: University of Illinois Press. pp. 271–317.

Suvin, D., 1979. Estrangement and cognition. In: D. Suvin, 1979. *Metamorphoses of science fiction*. New Haven, CT: Yale University Press. pp. 3–16.

The Terminator, 1984. [Film] Directed by James Cameron. USA: Orion Pictures Corporation.

Vint, S., 2012. Archaeologies of the 'amodern': science and society in Galileo's dream. *Configurations*, 20(1–2), (Winter–Spring 2012), pp. 29–51.

10 Shaping new horizons

Proactionary attitudes, precautionary principles and the experimentalities of science in society

Matthias Gross

Introduction

This chapter builds on a paradox. On the one hand, public and political expectations targeted at science, technology and engineering call for more safety, certainty or precaution. On the other hand, empirical research delivers insights on real-world decision-making and everyday practices that increasingly uncover how actors creatively cope with unavoidable uncertainty and ignorance and thus circumvent conventional approaches to risk assessments or classical evidence-based decision-analysis. This appears to be done since waiting for proper knowledge often leads to standstills and thus may preclude novel forms of imagination and opening horizons towards alternative futures.

In order to frame possible avenues towards shaping novel uses of imagination at the intersection of science and society in the face of increasing uncertainty and ignorance (cf. Elliot and Dickson, 2011), I will use the notion of real-world experiments to focus on the meeting point between science and society. By overturning the critical observation that in modern science there is an increasing tendency to unload risky research processes from the laboratory into the wider society, it is instead more productive to suggest that proactively shaping the future of science in society needs to be understood as part of a longer-term societal experimental learning process. Of crucial importance here is the experimental acknowledgement of ignorance as a basis for opening new horizons towards opportunities in twenty-first-century science and society. This, however, also raises new questions on the attribution of responsibilities, ethical concerns and the organization of scientific knowledge production in general. Ivo van de Poel (2016) has argued that experimental strategies are crucial once operational experience is scarce so that social benefits and risks cannot be assessed easily on the basis of laboratory experiments or computer simulations. Since, by definition, all innovations are moves into the unknown, novel technologies always come in tandem with new risks and side effects that often can only be coped with once the technologies are introduced and thus 'tested' in society.

In this chapter I will begin by explaining the notion of experimental strategies, positioning these within recent debates about a 'proactionary principle', a complementary viewpoint to the so-called precautionary principle that concludes that actions under situations of uncertainty and ignorance gaps should be postponed if not generally stopped. The proactionary principle (cf. More, 2005) focuses on the potential of innovative technologies and first and foremost on the opportunities – not only the negative side effects. The aim is to foster freedom to experiment and innovate even if risks are known. The precautionary principle, in turn, has been criticized a lot and, indeed, taken at full strength it would lead to a standstill since it can be used to justify using almost anything as a stopping rule. Anthony Giddens (2009) has prominently made the case that the precautionary principle can give both good guidance for mitigating climate change but also to support inaction if economic growth or the creation of jobs is the 'precautionary' goal. In a similar vein, Cass Sunstein (2005) has argued that the precautionary principle creates new risks of its own since waiting for proper knowledge (even if that was possible at all) can be risky in itself, sometimes even more so than acting under situations of uncertainty and even ignorance. Bruno Latour (2011, p. 12) put it perhaps most succinctly:

> If the precautionary principle meant the absurd idea that we should abstain to move until absolute certainty is reached, then that would be the end of European creativity, the end of science and technology, the end of all the collective experiments.

Whereas experiments comprise openness to the unknown by fostering unexpected results, but also a quick control of the unexpected element in order to learn from it, precautionary action appears to be in opposition to this.

In contrast to approaches that externalize ignorance by analysing risk assessments or that sidestep it by using rhetorics of certainty, I will suggest how the unexpected aspects of knowledge production can be described using the notion of experimentality. I use ignorance, or, as I will call it here, *nonknowledge*, as a central aspect of knowledge making and application, in order to knit it together with the idea of everyday experimentation.

After all, as Latour (2011, pp. 12–13) has stated, 'according to me, the precautionary principle … is a call for experimentation, invention, exploration, and of course risk-taking'. And, he continues,

> Far from waiting for absolute certainty before moving the little finger, we know we have to experiment and distribute equally audacity and what in German is called so beautifully *Sorge* and what we call in French *le souci*. Care and caution go together with risk-taking.

What Latour refers to here is a view to carefully coordinate and perhaps modularize openness to 'proactionarily' move into the unknown, hand in hand with the best precautionary measures available. Along this tension between precaution and proactionary strategies, in the following I will present some conceptual framing about the experimental society in order to discuss two case studies: the restoration and sanitation of contaminated industrial sites as well as the search for new energy below the ground – geothermal heat. In order to examine the notion of screening or even shaping new horizons and thus both opportunities and risks more closely and on equal footing, I will focus especially on situations of ignorance faced by practitioners and stakeholders in engineering, framing their strategies for moving forward despite not knowing – a form of experimental practice that seems increasingly normal and perhaps appropriate in twenty-first-century society. In the final section, I will reflect on how, in a democratically organized society, real-world experiments can be a basis for benefits for as many people as possible so that proactionary strategies can be turned into a socially responsible and feasible imperative in the twenty-first century.

Proaction + precaution = experimental strategies

Experiments and sudden social changes share some crucial similarities. An experiment in the most general sense can be defined as a cautiously observed venture into the unknown. An experiment is deliberately arranged to generate unexpected events; the surprising effects derived from the experimental set-up can be seen as the driver behind the production of new knowledge. Hans-Jörg Rheinberger (1997) has even stated that experiments have to engender unexpected events. And indeed, unexpected events lead to surprises, which help scientists become aware of what they did not know. The exact specification of what is not known can be rendered a hypothesis. To put it in more general terms, sudden unexpected changes that make the experimenters aware of their own ignorance (a falsified hypothesis) provide the impetus for new knowledge. The difference is that such abruptness is welcomed in a laboratory experiment, whereas in the everyday world of modern society it normally is not (Gross, 2010). This is important, given that calls for greater certainty and safety are being heard ever more loudly in contemporary political debate; at the same time, empirical research on decision-making can be seen as indicative of an increase in 'experimentalities' and as a reaction to unavoidable uncertainty and ignorance, the reverse side of modernist beliefs in scientific certainty. My observation is that experimental strategies are common among practitioners and engineers in many fields. Thus, perhaps the experimental strategies observed in real-world processes are indicative of a change in practical ways of coping with uncertainty and ignorance, even though the official rhetorics still suggest otherwise. Referring to contemporary initiatives in science and research as 'experiments' should thus be understood as a cautionary reminder that decision-making and interventions are never completely reliable. In this sense, experimentality is an appropriate term to refer to the realities within which decision-making and policy strategies operate. After all, everybody knows that in everyday life a whole host of things may happen unexpectedly. However, the theoretical tools available for framing what everybody seems to know are only in their infancy.

The proactionary principle has been formulated by futurist Max More, who sees innovation as crucial and highly valuable for modern societies, even if their development includes new risks (More, 2005). Thus, the proactionary principle should be understood in opposition to the precautionary principle. Sociologists Steve Fuller and Veronika Lipinska (2014) have taken up this view in order to reflect on what this could mean in a socially responsible society. Following these authors a major difference between humans and animals is that humans do not obey evolution but they actually understand it. This opens up opportunities for humans that should go beyond protection. Protecting people is often very important but if this becomes the main normative principle then humans become a type of endangered species that needs protection. Fuller and Lipinska then say that an important issue is that risks and taking risks need to be seen as something positive. This is a difficult point since risks 'themselves' can hardly be conceptualized as something positive. In everyday language, we may talk about positive risks (e.g. reflected in slogans such as 'no risk, no fun' or 'no pain, no gain') but the idea of taking risks in order to gain something does not necessarily make risk something positive. The fact that positive outcomes often require sacrifices is not the same as saying that these sacrifices are a good thing in themselves. One accepts more costs/risks only if one is hoping to get something else that is rewarding. Moreover, in many cases the positive aspects only become visible post hoc, thus no intention or goal-oriented activity to take risks can be meaningfully ascribed to any actor. Thus understood, risk thus should not be rendered something positive beforehand. Thus, if proactionary attitudes in cooperation with the precautionary principle should work well, traditional notions of risk simply defined or claimed as being positive may not work facing a concerned public.

The pivotal aim of concepts of risk developed from the second half of the twentieth century onwards has increasingly been to make the future predictable by means of probability calculations, thereby rendering it amenable to human intervention and planning. Understood thus, risk can be seen as a way of rationalizing an uncertain future. In these terms, dealing with ignorance clearly differs from taking or limiting risks, since the risk of a certain event occurring presupposes knowledge of both the character of events that may occur and the probability that they will do so.[1] In view of the above, I would add that in many cases when either scientists or lay people talk about risks it would be empirically and theoretically more useful and meaningful to frame decisions, utterances and practices according to different shadings of ignorance. In this sense the precautionary principle and the proactionary principle both share the same view. Both (at least implicitly) agree that in many cases risk assessments cannot deliver enough results to be used for decision-making. In this line of thinking, we can also talk about 'precautionary experiments' if no other action seems to be feasible. It is at this point where the precautionary principle (cf. de Sadeleer, 2007) and the proactionary imperative (cf. Fuller and Lipinska, 2014) can coincide. In other words, when there are no possibilities for predicting social consequences and only well-defined areas of the unknown (nonknowledge) is available in the form hypothesis-building or in terms of horizon-scanning, the proactionary measures as forms of precaution appears to be a feasible way forward. After all, the proactionary manifesto has been criticized, rightly so, that the possibilities for compensating victims that suffered for the advancement of humanity by supporting proactionary measures have not been developed at all (cf. Hauskeller, 2016).

The similarity in concrete cases between the proactionary and precautionary principle is based, as for instance Luigi Pellizzoni (2015) has observed, on the fact that both perspectives are united by the assumption of the impossibility to clearly predict or control social-technical interventions into the natural and social world. If the precautionary principle is seen as a tool to facilitate acting in spite of gaps of knowledge, as some proponents have done (cf. Myers and Raffensperger, 2006), and to learn from (small) failures in the absence of scientific proofs, then there may be cases where precaution actually means proactionary strategies – and vice versa. It is this overlap that I am interested in,[2] since for an increasing number of contemporary environmental problems it is impossible to assess and quantify the 'harm' that can be expected from a technology or an intervention into an ecosystem. Indeed, it is more often things that are not known that are most important for decision makers and thus more important for analysis. In this line of thinking, efforts to harness and to deploy ignorance are increasingly seen as important and profitable strategies within economic and political institutions. This position thus departs from the still common view that sees ignorance as necessarily detrimental, pointing instead to the broader challenges and (in some cases) the merits of not knowing and to the ways in which ignorance can even serve as a productive resource (cf. Littlewood, 2007; McGoey, 2012; Wehling, 2015).

Taking this line of thought further, in order to be able to act in an experimental setting the actors involved need to agree on what is not known and to take this into account for future planning. They need to decide to act in spite of (sometimes) well-defined ignorance, or what has been called 'nonknowledge' (Gross, 2010). Ignorance as nonknowledge refers to the acknowledgement that some things are unknown but are not specified enough to enable action. In the cases discussed in this, we most often observe a strategic regulation between, on the one hand, active or 'positive' nonknowledge, where the unknown is specified enough to be used for further planning and activity and, on the other hand, passive or 'negative' nonknowledge where the unknown is perhaps specified but is rendered unimportant or even dangerous to be acted upon at this point in time. Thus understood, nonknowledge should not generally be understood as ignorance, unawareness or the mere absence of knowledge but rather as a specific kind of expertise about what is not known. The aim is thus to specify ignorance so that it can be used in a meaningful and constructive way. Whereas having faith in total control and complete knowledge of ecological systems and social processes implies an ability to act only when everything is known in advance, an 'experimental approach' makes it possible to accommodate different factors despite the presence of gradations of the unknown.

From the experimenting to the experimental society

The idea to conceptualize processes outside the laboratory as experiments goes back to Francis Bacon's reflections on the relationship between the experimental method and society. On the one hand, Bacon was highly influential in shaping the world view that sees the experimenter's realm as distinct from the world of objects experimented upon and that privileges (rational) human beings over all others as masters of a world to which they essentially do not belong. Moreover, in Bacon's natural philosophy, only well-trained scientists in scientific institutions can access the secrets of nature (cf. Krohn, 2013). At the same time, Bacon's most provocative proposal was the idea that approval of the experimental method in science by the dignitaries of the state would turn society itself into a large-scale experiment (cf. Gross, 2016b). In this view, modern society should give the experimental method a chance since the promise of gains reaped by modern science cannot be backed up by anticipatory arguments but only by practising and implementing the new method in the larger society. There is good reason to buy into this Baconian argument. Ideally, the laboratory confines the problems and risks of research within its four walls, away from the wider society. Real-world experimentation at least partially unloads some of the uncertainties of research onto social and ecological systems. This being the case, it is possible to regard, for instance, ecological restoration processes outside the laboratory, for example as experiments – a view that poses a clear challenge to the premises of ecological (scientific) predictability, precaution and certainty.

Some classical notions of social experiments can be found in nineteenth-century thinkers' writings ranging from August Comte to John Stuart Mill (Brown, 1997), Jane Addams' Hull House (Gross, 2009) and the 1920s Chicago School of Sociology (Gross, 2010). Others trace the origin of current debates to John Dewey (Bogusz, 2013) or later authors such as Martin and Schinzinger (1996). The term and concept of 'experimenting society', however, was developed by Donald Campbell (1917–1996) and associates from the early 1960s onwards (see Campbell, 1998). Departing from this notion, and as a critical commentary on the way research processes and their related hazards were increasingly (and often unintentionally) extended beyond the limits of the laboratory into wider society during the 1980s, Wolfgang Krohn and Johannes Weyer (1989) outlined an understanding of scientific research that sees it as increasingly erasing the boundaries between the laboratory and wider society. Taking this rather critical stance as a point of departure and developing it further, twenty-first-century debates on real-world experiments (Gross and Hoffmann-Riem, 2005) have attempted to frame experimental practices as potentially useful strategies for coping with unavoidable uncertainty. If it is true that society has become a laboratory, then we may want to look for examples of how to deal successfully with the issues that arise in this extended laboratory. This calls for a notion of the experimental society that is based on a conceptualization of social processes as experimental endeavours undertaken to cope with the structural complexity and surprising dynamics of modern social life.

An experimental approach in this sense can be conceived of as a way of coordinating the contingent activities of diverse actors and enabling them to continue without interruption despite an acknowledged awareness of ignorance. In the view presented here, today's experimentality in modern societies did not emerge from social scientists' recommendations in the service of society and government but is understood as an indicator of societal strategies aimed at dealing proactively with unavoidable uncertainties, surprising events, ignorance and situation-specific experience. One of the central observations is that increasingly successful policymaking and regulation rests on 'experimental' frameworks that give actors wide-ranging freedoms to try out novel innovations, modes of implementation and legal regulations which, in turn, need to be evaluated so that the results can be compared with other experimental settings in order to make learning possible (cf. Felt *et al.*, 2007; Sabel and Zeitlin, 2010; van de Poel, 2016). Whereas most interpretations of the precautionary principle try to delay action before proof or at best to think about preventive actions in face of uncertainties, it does not spell out what should be done to move forward in the face of unavoidable unknowns. A truly experimental approach needs to be understood as one that knowingly moves into the unknown by trying to specify what is unknown (e.g. hypothesis-building) as clearly as possible and then using unavoidable failures or mishaps as a basis for learning measures. Consider an extreme example: In the case of geoengineering, for instance, some authors have reflected on whether it is indeed possible, due to time–space distanciation and indelible ignorance, to enter into the experiment democratically, and what this implies for how one should open up questions of governance at an early stage. Macnaghten and Szerszynski (2013) have shown that acceptability would only be expected under very specific conditions that are rarely found in real-world situations. Others, such as Corner, Pidgeon and Parkhill (2012), however, have stressed that research on public and stakeholder views in upstream engagement on geoengineering has shown that if participants are being facilitated to deliberate on the 'imaginaries' of the scientists, engineers and policymakers involved, then support may be possible despite unavoidable uncertainties and ignorance. Thus, if it is possible in a case such as geoengineering, then it should be possible in the relatively small-scale cases I discuss in this chapter.

If we now agree that experimentality in various forms is already part of modern society, what could this potentially mean for fields such as ecological restoration, landscape development or the utilization of renewable energy sources? Using some results from research on former industrial sites and drilling for new energy below the ground, the next two sections offer case studies of cleaning up contaminated industrial sites and drilling operations in geothermal energy utilization as part of the German energy transition towards renewable energy systems.

The impossibility of avoiding ignorance: experimenting with contaminants

Some 20,000 contaminated sites in Europe can be classified as so-called megasites owing to the complexity of soil and groundwater contamination found on them.[3] These megasites pose a range of tricky technical, social and sustainable management challenges (Bartke and Schwarze, 2015; Rizzo *et al.*, 2015). Some of the challenges relate to having to cope continuously with ignorance in the face of multiple sources of contaminants and plumes from previous industrial activities. Since much of the contamination on deserted brownfield sites originates from industrial production in the early twentieth century, records of accidents and waste disposal leading to contamination are often rare to non-existent. Very little documentation (e.g. archive materials or maps) of these chemicals dumped into the ground exists today. Worse still, the buffering capacities of many soil types and their ability to filter chemicals mean that contaminants are often not perceived until the damage is far advanced.

Given the predetermined time frames of most clean-up and restoration projects and the limited budget available, actors nowadays are usually aware that they will have to make decisions based on nonknowledge. They realize that it is useless to try to know things that cannot be known yet. To wait for proper knowledge would mean extending operations beyond project time frames and frittering away the chance to promote new investments and further economic development. However, this means that the actors on the ground have to expect that surprising things might happen. As soon as construction workers remove the topsoil, it is possible that they will find something that runs contrary to their expectations. The actors involved, then, learn to know in advance what they do not know (positive nonknowledge) and are able to use this as a basis for further planning and action. This positive nonknowledge can emerge from a general state of ignorance. Based on this nonknowledge, for instance, planners will consult engineering companies (because now it is known what is unknown) that will then take samples and evaluate the soil; they may subsequently conclude that an area is heavily contaminated with, say, liquid tar and various solvents. Commenting on such an incident, the representative from the engineering company in charge of a major revitalization project on contaminated land in the eastern German state of Saxony-Anhalt stated: 'The discovery of tar meant that the whole philosophy of the project had to be rethought. A completely new plan had to be drafted ad hoc.' In a similar vein, a representative from the project management explained: 'Construction and clean-up has to be understood as a continuous process. They [the workers on the ground] started to remove the material, and when they did so they found that the subsoil was totally different from what we expected.' For many stakeholders, this means that they have to act first and then later get the official permits for doing so. The actors thus involved often had to act in an ad hoc manner and, in so doing, move into a legal grey area. This phenomenon is not unusual and it shows that the impossibility of knowing everything was taken seriously in that it was turned into a constructive strategy. This is an important point. Doing otherwise would have slowed down the overall process significantly and would perhaps even have brought the project to a halt if the normal legal procedures for investigating mistakes had been triggered. In the case discussed here, the actors were prepared for unexpected events and so were able to make decisions quickly and flexibly. In other words, we can say that the actors involved agreed on what was not known and took this into account for their subsequent activities; that is, they decided to act on the basis of their (positive) nonknowledge. Positive nonknowledge, however, can be developed and 'shaped' into new knowledge to be fed into the next stage of the real-world experiment in order to subject it to further observation and assessment. Experts' willingness to disclose the limits of knowledge in their communication with other stakeholders can thus be understood as an important part of implementing successful landscape design and clean-up processes. This entails altering permits (administration), changing plans (engineering companies and research institutes) and, last but not least, being flexible in redeploying capital for remediation.

A further important issue with regard to successfully coordinating such projects is the institutionalization of contacts and information exchange. Although this might seem obvious at first sight, regular consultation of all institutions and actors involved for the purpose of exchanging information, discussing new developments and agreeing on strategies for adapting to new situations is not always standard procedure. Consultation also implies that all the actors involved must communicate their own nonknowledge, understood not in terms of failure or sloppy investigation but rather as a normal way of dealing with contaminants. The approach taken to cope with this phenomenon is one that accommodates project activities despite the presence of unknown factors.

Put another way, dealing with ignorance is not simply a process of trial and error or of learning from failure, as failures suggest that mistakes have been made. Instead, stakeholders and actors in the restoration of industrially contaminated land discussed in this chapter take seriously the impossibility of avoiding ignorance, so that there is no target for blame and no 'finger pointing'. The activities in contaminated site management presented here seem to include ignorance and nonknowledge explicitly as part of their forward planning. This seems to encourage the development of innovative strategies and to make full use of the potential and resources of the actors involved to achieve a common goal. New knowledge may emerge from these 'failures', but they open up new horizons for further inquiry and planning. Unlike probabilistic risk assessments, which often gloss over the unknown, 'successful failures' can help to specify ignorance so that the limits and the borders of knowing are intentionally taken into account for acting or planning. Given the enormous number of unknown factors involved in the dynamic layers of contaminated landscapes, acting in spite of nonknowledge can be seen as a crucial factor in our understanding of the experimentalities of landscape restoration.

New energy from below the ground

The extrusion of lava from volcanoes reminds us in a rather dramatic way that an enormous amount of pressure exists in the Earth's interior. Indeed, there is also a constant flow of heat that in theory can be tapped from literally any part of the Earth's surface. Thus, as a source of energy, heat from the Earth appears to have great potential for satisfying the growing energy needs of modern societies.[4] In general, geothermal heat (from the Greek *geo*, Earth, and *therme*, heat) is energy derived from the natural heat below the Earth's crust. Only 20 per cent of geothermal heat originates from the time of the original formation of the planet; 80 per cent stems from the *ongoing* radioactive decay of minerals, which is why geothermal energy can be considered a renewable energy. Unlike processes of coal or gas extraction, the energy removed from the source is continuously replaced by natural processes. Geothermal energy has come to occupy a special position among sources of renewable energy, as it is available all year round and at every hour of the night and day. In areas near tectonic plate boundaries (such as Hawaii, Iceland and northern Italy), geothermal energy has been in use for about a century. Currently, the largest producers of geothermal energy are the USA, the Philippines, Indonesia, Mexico and Italy. Some experts even claim that deep geothermal heat is the most promising option for replacing base-load energy produced up to now by nuclear power (Bauer *et al.*, 2014). However, although geothermal energy appears to offer great potential for future energy supplies, like all new technologies it also harbours major technical uncertainties and risks. In some regions, this has led to public distrust for this technology (cf. Kunze and Hertel, 2015; Stauffacher *et al.*, 2015). Major problems include the risks associated with exploration to find heat sources, unknowns connected with the stability of heat sources, and the lack of experience regarding long-term operations. The geotechnical risks, including earthquakes, are due in part to the lack of data available for developing useful models and monitoring systems. Thus, coping with knowledge gaps is the norm.

Some of the above-mentioned uncertainties in geothermal energy tapping are especially prevalent when it comes to probing downward into geothermal reservoirs up to five kilometres below the Earth's crust using new drilling technologies. Moving forward and downward in geothermal engineering can only be done under conditions of great uncertainty, where intuition and 'artistic' skills may be as important as so-called scientific rigor (cf. Gross, 2015). Furthermore, unlike any other energy source, geothermal energy is different because at the beginning of a geothermal power project the investment costs are the main source of unknowns. After the first stage of exploration (often geological studies conducted by the company involved but most often data available from the literature) the 'relevant developer ... will have to decide whether to continue developing the project or not and whether to *assume the risks* of the next phase'. (Bloomquist, Lund and Gehringer, 2013, p. 253). Not only does the project developer have to take the risks, but it also becomes clear that at this point anything (or nothing) can happen. Nevertheless, decisions need to be taken, even in the face of ignorance, because, as Bloomquist *et al.* (2013, p. 266) note, 'Computer models that accurately take into account this source of uncertainty to quantify the trade-offs with competing resources have yet to be developed'. Furthermore, drilling engineers we interviewed in our research[5] often summarized some of the above observations as follows: '[a]s novel options for resource use in the underground come up, novel risks have to be expected that are either not fully understood or are as yet unknown'. Another engineer said, 'There are not enough pre-explorations of local specificities available, but we need to move forward nevertheless.' This indicates that even engineers in favour of geothermal energy production point explicitly to the limits of knowing when dealing with geothermal energy. Thus, it appears that knowledge gaps and risks do not undermine optimism – quite the contrary. As in the case above concerning contaminated sites management, engineers agree on what was not known in order to take this into account for clarifying what is at stake (new horizons through positive nonknowledge) and for using this to shape new knowledge.

Despite many uncertainties and unknowns, which include the contamination of groundwater through drilling, managing risks and long-term liabilities, the prospect of achieving certainty through research on new remote control technologies and simulation tools for use in the geological substrata is still on the rise (cf. Mukuhira *et al.*, 2013). Critical voices, however, point to the fact that a major part of the problem with trying to tap the Earth's internal heat is the poorly understood geology. This appears to be the biggest obstacle for many geothermal engineers (cf. Mignan *et al.*, 2015). Given the unavoidable uncertainties in establishing alternative energy systems, decision-making in processes of tapping geothermal energy sources and related areas necessitate an open acknowledgement that nonknowledge is an inevitable given. One consequence of this insight would be to undermine the notion that mistakes and failures are necessarily always based on faulty decision-making or other human errors. Mistakes would take on a different meaning. This, of course, raises many unresolved questions. These include: when does nonknowledge as an explanation serve as subterfuge and when is it a legitimate way of dealing with the unknown? Problems such as these are obviously unsolved today. However, to acknowledge the experimental nature of tapping geothermal energy and, in doing so, to conceptualize the practices of actors in new ways appears to offer exciting options in our quest to understand the widely proclaimed energy transition in many countries.

Outlook: new horizons through real-world experiments

When new knowledge is acquired, it often leads to the perception of new horizons about what is *not* known. When it comes to planning and handling processes of restoring or revitalizing industrially altered landscapes, however, it is still common for both policymakers and scientists to officially state that decisions are based on reliable scientific knowledge ('the facts'). Only these known and observable 'facts' are included explicitly in policy and risk assessment deliberations. This leads to a gap between official rhetoric and the situation of practitioners on the ground for whom dealing experimentally with ignorance almost becomes the norm.

Furthermore, if it appears meaningful to characterize contemporary modern societies as leading towards more experimentality, and if experimentality is becoming a defining character of twenty-first-century society (cf. Gross, 2010; Hodson and Marvin, 2016; van de Poel, 2016), then public experimentation and the acknowledgement of nonknowledge are set to become more important than ever before. It would mean moving away from an orientation towards previous experience and historical extrapolations in nature, technology, and society (e.g. risk assessments) and shifting instead towards prospective and temporary notions of knowns and unknowns. This may be in line with what has been termed the proactionary imperative, but it harbours a few challenges that need to be addressed: the actors involved need to agree that no other means are reasonably available (as regards time, funding, and results to be expected) to gain knowledge about a certain issue that seems desirable (cf. Gross and Hoffmann-Riem, 2005). Although this will not always be possible, possibilities to stop the experiment need to be clarified beforehand. Furthermore, and this may be the most pressing and problematic issue, a fair distribution of benefits and hazards needs to be addressed (cf. Fuller and Lipinska, 2014, pp. 3–5; Robaey and Simons, 2015; van de Poel, 2016). In general, a way forward could be that a community that supports the decision to act in the face of ignorance and chooses an experimental path to move forward should be rewarded through compensation or tax reduction (cf. Duncan, 1999; Gross, 2010, p. 113). However, this is easier said than done. In a democratically anchored version of real-world experiments, this also means that as many people as possible need to benefit from it. Again, this may sound feasible on paper, but, as research on public attitudes and acceptability in renewable energy systems has shown (cf. Butler *et al.*, 2015), a wide range of different framings are needed to offer at least a basis for better understanding potential benefits. The task, then, becomes one of iteratively (or, in our terminology, experimentally) reformulating different understandings so that public perspectives are incorporated at the outset and form part of the imagined pathways. Whereas the precautionary principle creates a default negative expectation towards science and technology or even as an enemy until proven otherwise, an experimental approach will help to open up horizons for society so that many people can benefit even from 'adventurous' innovations.

Despite the discourse on European experimentalist governance (Sabel and Zeitlin, 2010), official rhetorics call for more certainty and more safety measures. After all, the precautionary principle is well established in everyday thinking. Decision-making based on nonknowledge would then be regarded as irresponsible tinkering – which it can certainly sometimes be, but to at least an equal measure it appears as an inevitable fact that should not be excluded from official statements. Despite this, real-world decision-making in a strictly regulated bureaucracy seems, quite paradoxically, to foster a positive engagement with ignorance and surprise, because in clearly defined spatial and temporal settings the experimental activities can be rendered an accepted part of an adversarial society, since they define legal spaces for failure, mishaps and risk-taking. Consider an example from the case of contaminated sites management above. One major factor which helped to generate an experimental strategy was an explicit agreement on shared objectives made at an *early* stage of the project. In addition to personal acknowledgements of what was not known, nonknowledge was even expressed in the contracts made between the actors. Special clauses in these contracts and permits – so-called collateral clauses – specifically mention the unknown. Collateral clauses are agreements between joint contractors to pool their guarantees in handling a large project. The objectives are to provide others with recourse to the contractor's warranty, to continue the project in the face of unexpected changes. These strategies were developed specifically for this site's remediation process. Thus, in this sense, each real-world experiment also becomes a pilot project. In this context, actors are able to harness the great potential entailed in moving forward along an experimental path.

This, of course, may also invite corporations or individual actors with unsavoury motives to use this as an excuse and to ensure the ability to change course if initial actions prove disastrous. This may be a reason why experimental strategies are mostly found in innovation and engineering projects such as the ones discussed in this chapter. In this sense, experimentality is no romantic idea of incrementalism, of muddling through or of some type of adaptive management where actors passively adapt to changing external conditions ('roll along with the punches'); instead, it demonstrates in a stark manner that dealing with unknowns can become a reality not by choice but by necessity for the sustainable future of a region. One is tempted to speculate that at a local or regional level experimental strategies may thrive best when there are strict state regulations in place that provide the 'proactionary' freedom to experiment and also to 'fail'. However, despite recent debates on urban and living laboratories (cf. Filho, 2012; Evans *et al.*, 2015; Göbel, 2015; Hodson and Marvin, 2016), experimentation and even less so 'failure' are not concepts that are generically held as positive ones in the language and minds of decision makers, stakeholders, concerned citizens and political leaders. We therefore have to expect more barriers to experimental 'horizon-scanning' in policymaking.

In the end, an important prerequisite for successfully failing through experimentation is that the actors involved are prepared to make decisions despite the existence of ignorance. It also needs public debate and the existence of citizens prepared to transformation and disruption. This includes a culture of acknowledging nonknowledge, of being against blame and finger pointing, and of clarifying the benefits of moving 'proactionarily' into the unknown. In official rhetoric this may be hard to find, but, as the cases discussed in this chapter should indicate, in different cultures of practice they seem to be alive and kicking.

Notes

1 Concepts of risk in many social and engineering sciences range from traditional actuarial analysis, classical probabilistic risk assessments (e.g. fault and event trees, scenarios and model building) to different psychological and sociological approaches. For the most updated overview on many fields of risk studies, see Burgess, Alemanno and Zinn (2016).

2 I stress this since a lot of other aspects in Fuller and Lipinska's book, such as the almost religious conviction that humans are to play God, coupled with a 'ruthless optimism' – as Hauskeller (2016) called it in comparing this with the accusation of Leibniz by Schopenhauer – appear highly problematic.
3 This section draws in part on Gross (2016a). The examples and interview quotes presented stem from different research projects on 'experimental strategies' in the revitalization of industrial sites in Germany, especially the SAFIRA II programme on remediation research in regionally contaminated aquifers, funded by the German Federal Ministry of Education and Research (BMBF) from 2006 to 2012. See also Gross and Bleicher (2013).
4 This section draws in part on Gross (2015). Further aspects of geothermal energy utilization can be found in Gross (2013), Bleicher and Gross (2015) and Gross and Mautz (2015).
5 This research was done as part of the *EnergyLandUse* 'integrated project' at Helmholtz Centre for Environmental Research (UFZ) in Leipzig, Germany. Further info on the project can be found at: www.ufz.de/index.php?en=32269.

References

Bartke, S. and Schwarze, R., 2015. No perfect tools: trade-offs of sustainability principles and user requirements in designing tools supporting land-use decisions between greenfields and brownfields. *Journal of Environmental Management*, 153, pp. 11–24.

Bauer, M., Freeden, W., Jacobi, H. and Neu, T. eds., 2014. *Handbuch Tiefe Geothermie: Prospektion, Exploration, Realisierung, Nutzung*. Berlin: Springer.

Bleicher, A. and Gross, M., 2015. User motivation, energy prosumers and regional diversity: sociological notes on using shallow geothermal energy. *Geothermal Energy*, 3(1), pp. 1–12.

Bloomquist, G., Lund, J. and Gehringer, M., 2013. Geothermal energy. In: G.M. Crawley, ed. 2013. *The world scientific handbook of energy: materials and energy – vol. 3*. Hackensack, NJ: World Scientific. pp. 245–273.

Bogusz, T., 2013. Experimentalismus statt Explanans? Zur Aktualität der pragmatistischen Forschungsphilosophie John Deweys. *Zeitschrift für Theoretische Soziologie*, 2(2), pp. 239–252.

Brown, R., 1997. Artificial experiments on society: Comte, C.G. Lewis and Mill. *Journal of Historical Sociology*, 10(1), pp. 74–97.

Burgess, A., Alemanno, A. and Zinn, J. eds., 2016. *Routledge handbook of risk studies*. London: Routledge.

Butler, C., Demski, C., Parkhill, K., Pidgeon, N. and Spence, A., 2015. Public values for energy futures: framing, indeterminacy and policy making. *Energy Policy*, 87, pp. 665–672.

Campbell, D.T., 1998. The experimenting society. In: W.N. Dunn, ed. 1998. *The experimenting society: essays in honor of Donald T. Campbell*. New Brunswick, NJ: Transaction. pp. 35–68.

Corner, A., Pidgeon, N. and Parkhill, K., 2012. Perceptions of geoengineering: public attitudes, stakeholder perspectives, and the challenge of 'upstream' engagement. *WIREs Climate Change*, 3(5), pp. 451–466.

de Sadeleer, N. ed., 2007. *Implementing the precautionary principle: approaches from the Nordic countries, EU and USA*. London: Earthscan.

Duncan, I.J., 1999. A community that accepts risk should be rewarded. *Risk. Decision & Policy*, 4(3), pp. 191–199.

Elliot, K.C. and Dickson, M., 2011. Distinguishing risk and uncertainty in risk assessments of emerging technologies. In: T.B. Zülsdorf, C. Coenen, A. Ferrari, U. Fiedeler, C. Milburn and M. Wienroth, eds. 2011. *Quantum engagements: social reflections of nanoscience and emerging technologies*. Heidelberg: AKA Verlag. pp. 165–176.

Evans, J., Jones, R., Karvonen, A., Millard, L. and Wendler, J., 2015. Living labs and co-production: university campuses as platforms for sustainability science. *Current Opinion in Environmental Sustainability*, 16 (October), pp. 1–6.

Felt, U., Wynne, B., Callon, M. Gonçalves, M.E., Jasanoff, S. *et al.*, 2007. *Taking European knowledge society seriously*. Luxembourg: Office for Official Publications of the European Communities.

Filho, W.L. ed., 2012. *Sustainable development at universities: new horizons*. Frankfurt am Main: Peter Lang.

Fuller, S. and Lipinska, V., 2014. *The proactionary imperative: a foundation for transhumanism*. Houndmills: Palgrave Macmillan.

Giddens, A., 2009. *The politics of climate change*. Oxford: Polity.

Göbel, H.K., 2015. *The re-use of urban ruins: atmospheric inquiries of the city*. London: Routledge.

Gross, M., 2009. Collaborative experiments: Jane Addams, Hull-house, and experimental social work. *Social Science Information*, 48(1), pp. 81–95.

Gross, M., 2010. *Ignorance and surprise: science, society, and ecological design*. Cambridge, MA: MIT Press.

Gross, M., 2013. Old science fiction, new inspiration: communicating unknowns in the utilization of geothermal energy. *Science Communication*, 35(6), pp. 810–818.

Gross, M., 2015. Journeying to the heat of the earth: from Jules Verne to present-day geothermal adventures. *Engineering Studies*, 7(1), pp. 28–46.

Gross, M., 2016a. Layered industrial sites: experimental landscapes and the virtues of ignorance. In: M. Hourdequin and D.G. Havlick, eds. 2016. *Restoring layered landscapes: history, ecology, and culture*. Oxford: Oxford University Press. pp. 73–91.

Gross, M., 2016b. Give me an experiment and I will raise a laboratory. *Science, Technology & Human Values*, 41(4), pp. 613–634.

Gross, M. and Bleicher, A., 2013. 'It's always dark in front of the pickaxe': organizing ignorance in the long term remediation of contaminated land. *Time & Society*, 22(3), pp. 316–334.

Gross, M. and Hoffmann-Riem, H., 2005. Ecological restoration as a real-world experiment: designing robust implementation strategies in an urban environment. *Public Understanding of Science*, 14(3), pp. 269–284.

Gross, M. and Mautz, R., 2015. *Renewable energies*. London: Routledge.

Hauskeller, M., 2016. Book review symposium on *The proactionary imperative*. *Sociology*, 50(3), pp. 612–614.

Hodson, M. and Marvin, S. eds., 2016. *Retrofitting cities: priorities, governance, and experimentation*. London: Routledge.

Krohn, W., 2013. Von der *Instauratio Magna* zur Dystopia Magna? Das endlose Experiment der gesellschaftlichen Modernisierung. In: V. Chilese and H.-P. Preußer, eds. 2013. *Technik in Dystopien*. Heidelberg: Universitätsverlag Winter. pp. 31–46.

Krohn, W. and Weyer, J., 1989. Gesellschaft als Labor: Die Erzeugung sozialer Risiken durch experimentelle Forschung. *Soziale Welt*, 40(3), pp. 349–373.

Kunze, C. and Hertel, M., 2015. Tiefe Geothermie: Von hohen Erwartungen zur Risikotechnologie. *Gaia*, 24(3), pp. 169–173.

Latour, B., 2011. From multiculturalism to multinaturalism: what rules of method for the new socio-scientific experiments? *Nature + Culture*, 6(1), pp. 1–17.

Littlewood, R. ed., 2007. *On knowing and not knowing in the anthropology of medicine*. Walnut Creek, CA: Left Coast Press.

Martin, M.W. and Schinzinger, R., 1996 [1983]. *Ethics in engineering*. New York, NY: McGraw Hill.

Macnaghten, P. and Szerszynski, B., 2013. Living the global social experiment: an analysis of public discourse on solar radiation management and its implications for governance. *Global Environmental Change*, 23(2), pp. 465–474.

McGoey, L., 2012. The logic of strategic ignorance. *British Journal of Sociology*, 63(3), pp. 553–576.

Mignan, A., Landtwing, D., Kästli, P., Mena, B. and Wiemer, S., 2015. Induced seismicity risk analysis of the 2006 Basel, Switzerland, enhanced geothermal system project: influence of uncertainties on risk mitigation. *Geotermics*, 53 (January), pp. 133–146.

More, M., 2005. *The proactionary principle*. Available at: www.maxmore.com/proactionary.html [Accessed 8 September 2015].

Mukuhira, Y., Asanuma, H., Niitsuma, H. and Häring, M.O., 2013. Characteristics of large-magnitude microseismic events recorded during and after stimulation of a geothermal reservoir at Basel, Switzerland. *Geothermics*, 45(1), pp. 1–17.

Myers, N.J. and Raffensperger, C. eds., 2006. *Precautionary tools for reshaping environmental policy*. Cambridge, MA: MIT Press.

Pellizzoni, L., 2015. *Ontological politics in a disposable world: the new mastery of nature*. Farnham: Ashgate.

Rheinberger, H.-J., 1997. *Toward a history of epistemic things: synthesizing proteins in the test tube*. Palo Alto, CA: Stanford University Press.

Rizzo, E., Pesce, M., Pizzol, L., Alexandrescu, F.M., Giubilato, E., Critto, A., Marcomini, A. and Bartke, S., 2015. Brownfield regeneration in Europe: identifying stakeholder perceptions, concerns, attitudes and information needs. *Land Use Policy*, 48, pp. 437–453.

Robaey, Z. and Simons, A., 2015. Responsible management of social experiments: challenges for policymaking. In: B.-J. Koops, I. Oosterlaken, H. Romijn, T. Swierstra and J. van den Hoven, eds. 2015. *Responsible innovation 2: concepts, approaches, and applications*. Heidelberg: Springer. pp. 87–103.

Sabel, C.F. and Zeitlin, J. eds., 2010. *Experimentalist governance in the European Union*. Oxford: Oxford University Press.

Stauffacher, M., Muggli, N., Scolobig, A. and Moser, C., 2015. Framing deep geothermal energy in the mass media: the case of Switzerland. *Technological Forecasting and Social Change*, 98, pp. 60–70.

Sunstein, C.R., 2005. *Laws of fear: beyond the precautionary principle*. Cambridge: Cambridge University Press.

van de Poel, I., 2016. An ethical framework for evaluating experimental technology. *Science and Engineering Ethics*, 22(3), pp. 667–686.

Wehling, P. ed., 2015. *Vom Nutzen des Nichtwissens: Sozial- und kulturwissenschaftliche Perspektiven*. Bielefeld: Transcript.

Index

Page numbers in *italics* denote tables, those in **bold** denote figures.